1 MONTH OF
FREE
READING

at

www.ForgottenBooks.com

By purchasing this book you are eligible for one month membership to ForgottenBooks.com, giving you unlimited access to our entire collection of over 1,000,000 titles via our web site and mobile apps.

To claim your free month visit:

www.forgottenbooks.com/free614713

ISBN 978-0-484-62951-5
PIBN 10614713

PROCEEDINGS

AND

TRANSACTIONS

OF THE

LIVERPOOL BIOLOGICAL SOCIETY.

VOL. X.

SESSION 1895-96.

LIVERPOOL:
PRINTED BY T. DOBB & CO., 229, BROWNLOW HILL.
1896.

CONTENTS.

I. PROCEEDINGS.

II. TRANSACTIONS.

PROCEEDINGS

OF THE

LIVERPOOL BIOLOGICAL SOCIETY.

OFFICE-BEARERS AND COUNCIL.

Ex-Presidents:

1886—87 Prof. W. MITCHELL BANKS, M.D., F.R.C.S
1887—88 J. J. DRYSDALE, M.D.
1888—89 Prof. W. A. HERDMAN, D.Sc., F.R.S.E.
1889—90 Prof. W. A. HERDMAN, D.Sc., F.R.S.E.
1890—91 T. J. MOORE, C.M.Z.S.
1891—92 T. J. MOORE, C.M.Z.S., A.L.S.
1892—93 ALFRED O. WALKER, J.P., F.L.S.
1893—94 JOHN NEWTON, M.R.C.S.
1894—95 Prof. F. GOTCH, M.A., F.R.S.

SESSION X., 1895–96.

President:

Prof. R. J. HARVEY GIBSON, M.A.

Vice-Presidents:

Prof. W. A. HERDMAN, D.Sc., F.R.S.
HENRY O. FORBES, LL.D., F.Z.S.

Hon. Treasurer:

ISAAC C. THOMPSON, F.L.S., F.R.M.S.

Hon. Librarian

ANDREW SCOTT.

Hon. Secretary

JOSEPH. A. CLUBB, B.Sc. (Vict.).

Council:

H. C. BEASLEY.
Prof. BOYCE, M.B., M.R.C.S.
W. J. HALLS.
J. SIBLEY HICKS, M.D., F.L.S.
Rev. T. S. LEA, M.A.
G. H. MORTON, F.G.S.

JOHN NEWTON, M.R.C.S.
T. C. RYLEY.
W. E. SHARP.
A. T. SMITH, Jun.
A. O. WALKER, F.L.S.
J. WIGLESWORTH, M.D.

REPORT of the COUNCIL.

DURING the Session 1895-96 there have been seven ordinary meetings of the Society, held as heretofore at University College. A field meeting was arranged for to Hilbre Island, but the weather proved so bad, that it had to be abandoned.

The communications made to the Society have been representative of almost all branches of Biology and many interesting exhibits have been submitted at the meetings.

The change made in the procedure whereby half an hour at the beginning of the meeting is devoted to miscellaneous exhibits, tea and conversation, has proved on the whole satisfactory.

As on former occasions the Society has been favoured with an address from a distinguished Biologist from another centre, viz., Dr. Scott, Honorary Keeper of the Jodrell Laboratory, Kew, whose paper, on "Ferns and Flowering Plants,—a chapter in evolution" was greatly appreciated.

The Library still continues to make satisfactory progress as shown by the Librarian's Report which follows.

The Treasurer's usual statement and balance sheet are appended.

No alterations have been made in the Laws of the Society during the past session.

Mr. Alfred Leicester, and Dr. Hanitsch former members of the Society, and both of whom have rendered valuable services to the Society have been added to the roll of Honorary Members.

The members at present on the roll are as follows :—

Honorary Members..........	9
Ordinary Members...........	66
Student Members............	20
Total	95

SUMMARY of PROCEEDINGS at the MEETINGS.

The first meeting of the tenth session was held at University College on Friday, 11th October, 1895.

1. The first part of the proceedings took place in the large Zoology Laboratory from 7 to 8 o'clock. Tea was served at one end of the room and on the tables were the following exhibits :—

    By Prof. Herdman : a series of sub-marine deposits from the Irish Sea.

    By Prof. Harvey Gibson ; some recent additions to the Botanical Museum of University College, from Kew Gardens.

    By Dr. H. O. Forbes ; a living specimen of the electric cat-fish, (*Malapterurus electricus*), from the Aquarium of the Free Public Museum.

    By J. A. Clubb, B.Sc. ; some living specimens of fresh-water Polyzoa under the microscope.

The President-elect (Prof. R. J. Harvey Gibson, M.A.,) took the chair at 8 o'clock, in the Zoology Theatre.

2. The Report of the Council on the Session 1894-95 (see "Proceedings," Vol. IX., p. viii.) was read and adopted.

3. The Treasurer's Balance Sheet for the Session 1894-95 (see "Proceedings," Vol. IX., p. xxxiv.) was submitted and approved.

4. The Librarian's Report (see "Proceedings," Vol. IX., p. xxv.) was submitted and approved.

5. The following Office-bearers and Council for the ensuing Session were elected :—Vice-Presidents, Prof. W. A. Herdman, D.Sc., F.R.S., H. O. Forbes, LL.D., F.Z.S. ; Hon. Treasurer, I. C. Thompson, F.L.S.,

F.R.M.S.; Hon. Librarian, Andrew Scott; Hon. Secretary, Joseph A. Clubb, B.Sc. (Vict.); Council, H. C. Beasley, Prof. Boyce, M.B., M.R.C.S., W. J. Halls, J. Sibley Hicks, M.D., Rev. T. S. Lea, M.A., G. H. Morton, F.G.S., John Newton, M.R.C.S., T. C. Ryley, W. E. Sharp, A. T. Smith, Jun., A. O. Walker, F.L.S., and J. Wiglesworth, M.D.

6. Mr. Alfred Leicester was elected an Honorary Member of the Society.

7. The President delivered the Inaugural Address, entitled "Botanic Gardens—past and present" (see "Transactions," p. 1). A vote of thanks proposed by Dr. Wiglesworth, seconded by Mr. Alf. O. Walker, was carried with acclamation.

———

The second meeting of the tenth session was held at University College on Friday, November 8th, 1895. The President in the chair.

1. The following exhibits were on view in the Zoological Laboratory from 7 to 8 o'clock :—

By Dr. Newton ; some examples of pond life under the microscope.

By Dr. Forbes ; specimens of marine animals preserved in formalin.

2. Mr. G. H. Morton, F.G.S., gave a short note on some plant remains from the Carboniferous Limestone, and exhibited some artistically prepared restorations drawn by Miss Wood.

3. Prof. Herdman, F.R.S., communicated the Ninth . Annual Report on the work of the Liverpool Marine Biology Committee, and the Port Erin Biological Station (see " Transactions," p. 34.)

———

The third meeting of the tenth session was held at

University College on Friday, December 13th, 1895. The President in the chair.

1. In the Zoological Laboratory, Mr. Andrew Scott exhibited drawings of eleven species of Copepoda new to the District; Prof. Harvey Gibson showed a series of Botanical preparations arranged for teaching purposes, and Mr. F. J. Cole exhibited microscopic specimens of the nerves of skate, etc.

2. Dr. Grossmann gave an intensely interesting account of Whaling off the Faröe Islands. By means of a beautiful series of lantern slides, the sighting, catching and cutting up of the big whales were most graphically portrayed. An interesting discussion ensued on certain points brought forward in the paper, in which Dr. Newton, Mr. Thompson, the Hon. Secretary and others took part.

The fourth meeting of the tenth session was held at University College on Friday, January 10th, 1896. The President in the chair.

1. In the Zoological Laboratory, Prof. Herdman exhibited a series of drawings and microscopic preparations illustrative of the anatomy of the oyster.

2. Mr. I. C. Thompson, F.L.S., gave an account of some free-swimming Copepoda from the West Coast of Ireland.

3. A joint interim Report on Green Oysters and Disease was submitted by Profs. Herdman and Boyce. A series of lantern slides illustrated some of the points of the paper.

4. Dr. H. O. Forbes laid on the table some notes on the brain of the Chimpanzee (*Anthropopithecus calvus*), which will be published in the next volume of Transactions, together with some additional points.

The fifth meeting of the tenth session was held at University College on Friday, February 14th, 1896. The President in the chair.

1. In the Zoology Laboratory, among other exhibits was a form of photographic camera adapted for taking photographs of living algæ in rock pools, and used by the Rev. T. S. Lea at Port Erin.

2. Rev. T. S. Lea gave an interesting exhibition of an excellent series of these photographs.

3. Dr. H. O. Forbes laid before the Society a paper on the evidence in favour of the former existence of an Antarctic Continent. He drew some forcible arguments from the distribution of both animal and plant life at present prevailing in the extremities of the existing continents of Australia, Africa and S. America, and mainly on the evidence of this distribution he showed an outline map of the possible land connections which may at one time have prevailed. An interesting discussion followed in which Prof. Herdman, Mr. Lomas, Mr. Fitzpatrick and others took part. Various points of Dr. Forbes, paper were freely criticised, and it was generally agreed that a greater knowledge of the depth of the sea in the area involved was required.

The sixth meeting of the tenth session was held in University College on Friday, March 13th, 1896. The Vice-President (Prof. Herdman) in the chair.

1. In the Zoology Laboratory, Mr. I. C. Thompson exhibited a form of incandescent gas light suitable as a microscopic lamp.

2. Mr. Alfred O. Walker contributed a paper on the proportion of genera to species in certain localities,

and advanced some interesting statistics in regard to the same. A discussion followed.

3. Prof. Herdman submitted the Annual Report on the work of the Sea-Fisheries Laboratory for 1895 by Mr. Andrew Scott and himself (see "Transactions," p. 103).

4. Prof. Herdman communicated a series of extracts translated by Mrs. Herdman from Dr. Johan Hjorts "Hydrografisk--Biologiske Studier over Norske Fisherier," a paper dealing with the currents and their effect on the Fisheries of the North Sea.

The seventh meeting of the tenth session was held in University College on Friday, May 8th, 1896. The President in the chair.

1. In the Zoological Laboratory a number of microscopic slides were exhibited.

2. Dr. Richard Hanitsch was elected an Honorary Member of the Society.

3. Dr. Scott, Honorary Keeper of the Jodrell Laboratory, Kew, gave an address on "Ferns and Flowering Plants—a chapter in evolution" (see Abstract "Transations," p. 181). On the motion of Prof. Weiss, seconded by Dr. Forbes, F.Z.S., and supported by Prof. Herdman, a hearty vote of thanks was accorded to Dr. Scott.

4. Mr. A. J. Ewart, B.Sc., Ph.D., contributed a paper on "Further observations on the Vitality of Seeds" (see "Transactions," p. 185).

The Annual Field Meeting was arranged for July 25th to Hilbre Island, but as rain was falling heavily at the time of the starting of the train, it was reluctantly decided to abandon the trip. An adjournment to the Museum was agreed to, for the purpose of holding a short business meeting.   Prof. Herdman took the chair.

On the motion of Prof. Herdman seconded by J. A. Clubb, B.Sc., Dr. Forbes, F.Z.S., was elected President for the next session.

LAWS of the LIVERPOOL BIOLOGICAL SOCIETY.

I.—The name of the Society shall be the "LIVERPOOL BIOLOGICAL SOCIETY," and its object the advancement of Biological Science.

II.—The Ordinary Meetings of the Society shall be held at University College, at Seven o'clock, during the six Winter Months, on the second Friday evening in every month, or at such other place or time as the Council may appoint.

III.—The business of the Society shall be conducted by a President, two Vice-Presidents, a Treasurer, a Secretary, a Librarian, and twelve other Members, who shall form a Council; four to constitute a quorum.

IV.—The President, Vice-Presidents, Treasurer, Secretary, Librarian, and Council shall be elected annually, by ballot, in the manner hereinafter mentioned.

V.—The President shall be elected by the Council (subject to the approval of the Society) at the last Meeting of the Session, and take office at the ensuing Annual Meeting.

VI.—The mode of election of the Vice-Presidents, Treasurer, Secretary, Librarian, and Council shall be in the form and manner following:—It shall be the duty of the retiring Council at their final meeting to suggest the names of Members to fill the offices of Vice-Presidents, Treasurer, Secretary, Librarian, and of four Members who were not

on the last Council to be on the Council for the ensuing
session, and formally to submit to the Society, for election
at the Annual Meeting, the names so suggested.   The
Secretary shall make out and send to each Member of the
Society, with the circular convening the Annual Meeting,
a printed list of the retiring Council, stating the date of
the election of each Member, and the number of his atten-
dances at the Council Meetings during the past session;
and another containing the names of the Members suggested for election, by which lists, and no others, the votes
shall be taken.   It shall, however, be open to any Member
to substitute any other names in place of those upon the
lists, sufficient space being left for that purpose.   Should
any list when delivered to the President contain other
than the proper number of names, that list and the votes
thereby given shall be absolutely void.   Every list must
be handed in personally by the Member at the time of
voting.   Vacancies occurring otherwise than by regular
annual retirement shall be filled by the Council.

VII.—Every Candidate for Membership shall be proposed by three or more Members, one of the proposers
from personal knowledge.   The nomination shall be read
from the Chair at any Ordinary Meeting, and the Candidate therein recommended shall be balloted for at the
succeeding Ordinary Meeting.   Ten black balls shall
exclude.

VIII.—When a person has been elected a Member, the
Secretary shall inform him thereof, by letter, and shall at
the same time forward him a copy of the Laws of the
Society.

IX.—Every person so elected shall within one calendar
month after the date of such election pay an Entrance Fee
of Half a Guinea and an Annual Subscription of one

Guinea (except in the case of Student Members); but the Council shall have the power in exceptional cases, of extending the period for such payment. No Entrance Fee shall be paid on re-election by any Member who has paid such fee.

X.—The Subscription (except in the case of Student Members) shall be One Guinea per annum, payable in advance, on the day of the Annual Meeting in October.

XI.—Members may compound for their Annual Subscriptions by a single payment of Ten Guineas.

XII.—There shall also be a class of Students Members, paying an Entrance Fee of Two Shillings and Sixpence, and a Subscription of Five Shillings per annum.

XIII.—All nominations of Student Members shall be passed by the Council previous to nomination at an Ordinary Meeting. When elected, Student Members shall be entitled to all the privileges of Ordinary Members, except that they shall not receive the publications of the Society, nor vote at the Meetings, nor serve on the Council.

XIV.—Resignation of Membership shall be signified *in writing* to the Secretary, but the Member so resigning shall be liable for the payment of his Annual Subscription, and all arrears up to date of his resignation.

XV.—The Annual Meeting shall be held on the second Friday in October, or such other convenient day in the month as the Council may appoint, when a Report of the Council on the affairs of the Society, and a Balance Sheet, duly signed by the Auditors previously appointed by the Council, shall be read.

XVI.—Any person (not resident within ten miles of Liverpool) eminent in Biological Science, or who may have rendered valuable services to the Society, shall be eligible as an Honorary Member; but the number of such Members shall not exceed fifteen at any one time.

XVII.—Captains of vessels and others contributing objects of interest shall be admissible as Associates for a period of three years, subject to re-election at the end that time.

XVIII.—Such Honorory Members and Associates shall be nominated by the Council, elected by a majority at an Ordinary Meeting, and have the privilege of attending and taking part in the Meetings of the Society, but not voting.

XIX.—Should there appear cause in the opinion of the Council for the expulsion from the Society of any Member, a Special General Meeting of the Society shall be called by the Council for that purpose; and if two-thirds of those voting agree that such Member be expelled, the Chairman shall declare this decision, and the name of such Member shall be erased from the books.

XX.—Every Member shall have the privilege of introducing one visitor at each Ordinary Meeting. The same person shall not be admissible more than twice during the same session.

XXL—Notices of all Ordinary or Special Meetings shall be issued to each Member by the Secretary, at least three days before such Meeting.

XXII.—The President, Council, or any ten Members can convene a Special General Meeting, to be called within fourteen days, by giving notice in writing to the Secretary, and stating the object of the desired Meeting. The circular convening the Meeting must state the purpose thereof.

XXIII.—Votes in all elections shall be taken by ballot, and in other cases by show of hands, unless a ballot be first demanded.

XXIV.—No alteration shall be made in these Laws, except at an Annual Meeting, or a Special Meeting called for that purpose; and notice in writing of any proposed alteration shall be given to the Council, and read at the Ordinary Meeting, at least a month previous to the meeting at which such alteration is to be considered, and the proposed alteration shall also be printed in the circular convening such meeting; but the Council shall have the power of enacting such Bye-Laws as may be deemed necessary, which Bye-Laws shall have the full power of Laws until the ensuing Annual Meeting, or a Special Meeting convened for their consideration.

BYE-LAW.

Student Members of the Society may be admitted as Ordinary Members without re-election upon payment of the Ordinary Member's Subscription; and they shall be exempt from the Ordinary Member's entrance fee.

LIST of MEMBERS of the LIVERPOOL BIOLOGICAL SOCIETY.

SESSION 1894-95.

A. ORDINARY MEMBERS.

(Life Members are marked with an asterisk.)

ELECTED.

1886 Banks, Prof. W. Mitchell, M.D., F.R.C.S., 28, Rodney-street

1890 Batters, E. A. L., B.A., LL.B., F.L.S., The Laurels, Wormly, Herts

1886 Barron, Prof. Alexander, M.B., M.R.C.S., 31, Rodney-street

1888 Beasley, Henry C., Prince Albert-road, Wavertree

1894 Boyce, Prof., University College, Liverpool

1889 Brown, Prof. J. Campbell, 27, Abercromby-square

1887 Caine, Nathaniel, Spital, Bromborough

1886 Caton, R., M.D., F.R.C.P., Lea Hall, Gateacre

1886 Clubb, J. A., B.Sc., HON. SECRETARY, Free Public Museum, Liverpool

1895 Cole, F. J., University College, Liverpool

1890 Davies, D., 55, Berkley-street

1891 Dismore, Miss, 65, Shewsbury-road, Oxton

1889 Dwerryhouse, A. R., 8, Livingstone-avenue, Sefton Park

1890 Ewart, A. J., B.Sc., 33, Berkley-street, Liverpool

1894 Forbes, H. O., LL.D., F.Z.S., VICE-PRESIDENT, Free Public Museum, Liverpool

1891 Garstang, W., M.A., Lincoln College, Oxford

1886 Glynn, Prof. T. R., M.D., F.R.C.P., 62, Rodney-street

1886   Gibson, Prof. R. J., M.A., **F.L.S.**, PRESIDENT, University College

1891   Gotch, Prof. F., F.R.S., Oxford

1894   Greening, Linnæus, **F.L.S.**, 5, Wilson Patten-street, Warrington

1894   Grossmann, Karl, M.D., 70, Rodney-street

1886   Halls, W. J., 35, Lord-street

1887   Healey, George F., 32, Croxteth-grove, Sefton Park

1886   Herdman, Prof. W. A., D.Sc., F.R.S., VICE-PRESIDENT, University College

1893   Herdman, Mrs., B.Sc., 32, Bentley-road, Liverpool

1891   Hicks, J. Sibley, M.D., 2, Erskine-street

1894   Hickson, Prof. S. J., Owens College, Manchester

1888   *Hurst, C. H., Ph.D., 41, Stephen's-green, Dublin

1886   Jones, Charles W., Field House, Prince Alfred-road, Wavertree

1894   Jones, Charles E., Elpie, Prenton-rd., W., B'head

1895   Klein, Rev. L. de Beaumont, D.Sc., **F.L.S.**, 6, Devonshire-road

1894   Lea, Rev. T. S., 3, Wellington Fields, Wavertree

1896   Laverock, W. S., M.A., B.Sc., Free Museum, Liverpool

1886   Lomas, J., Assoc.N.S.S., Salen, Amery Grove, Birkenhead

1893   Macdonald, J. S., B.A., Physiological Lab., Univ. College, Liverpool

1888   Melly, W. R., 90, Chatham-street

1886   McMillan, William S., **F.L.S.**, Brook-road, Maghull

1886   Morton, G. H., F.G.S., 209, Edge-lane, E.

1888   Newton, John, M.R.C.S., 44, Rodney-street

1887   Narramore, W., **F.L.S.**, 5, Geneva-road, Elm Park

1894   Paterson, Prof., M.D., M.R.C.S., University College, Liverpool

1894   Paul, Prof. F. T., Rodney-street, Liverpool
1891   Phillips, Miss F., 3, Green-lawn, Rock Ferry
1892   Phillips, E. J. M., L.D.S., M.R.C.S., Rodney-st.
1886   *Poole, Sir James, J.P., Abercromby Square
1890   Rathbone, Miss May, Backwood, Neston
1895   Ricketts, C., M.D., 1a, Cleveland-street, Birkenhead
1887   Robertson, Helenus R , Springhill, Church-road, Wavertree
1887   Ryley, Thomas C., 10, Waverley-road
1895   Sandeman, G., University College, Liverpool
1892   Sephton, Rev. J., M.A., 90, Huskisson-street
1894   Scott, Andrew, Hon. Librarian, University College, Liverpool
1895   Sherrington, Prof., M.D., F.R.S., University College, Liverpool
1886   Smith, Andrew T., Jun., 13, Bentley-road, Prince's Park
1895   Smith J., Rose Villa, Lachford, Warrington
1893   Tate, Francis, F.C.S., Hackins Hey, Liverpool
1886   Thompson, Isaac C., F.L.S., F.R.M.S., Hon. Treasurer, 53, Croxteth-road
1889   Thornely, Miss L. R., Baycliff, Woolton Hill
1888   Toll, J. M., Kirby Park, Kirby
1886   Walker, Alfred O., J.P., F.L.S., Colwyn Bay
1889   Williams, Miss Leonora, Hill Top, Bradfield, nr. Sheffield
1891   Wiglesworth, J., M.D., County Asylum, Rainhill
1891   Wood, G. W., F.I.C., Riggindale-road, Streatham, London
1892   Weiss, Prof., Owens College, Manchester
1892   Young, T. F., M.D., 12, Merton-road, Bootle

B. Student Members.

Armstrong, Miss A., 26, Trinity-road, Bootle

Burnet, A., University College, Liverpool
Crowther, H. P., University College, Liverpool
Chadwick, H. C., Free Museum, Bootle
Christophers, S. R., 10, Lily-road, Fairfield
Crompton, Miss C. A., University College, Liverpool
Depree, S. S., 3, Morley-road, Southport
Dickinson, T., 3, Clark-street, Prince's Park
Dumergue, A. F., 7, Montpellier-terrace, Up. Parliament-st.
Dutton, J. E., Kings-street, Rock Ferry
Ewart, R. J., 33, Berkley-street, Liverpool
Hannah, J. H. W., 61, Avondale-road, Sefton Park
Harvey, E. J. W., 5, Cairns-street, Liverpool
Henderson, W. S., B.Sc., 2, Holley-road, Fairfield
Hurter, D. G., Holly Lodge, Cressington
Linton, S. F., St. Pauls Vicarage, Clifton-road, Birkenhead
Lowe, O. W. A., 4, Wexford-road, Oxton
Quinby, F. G., 11, Belvidere-road, Liverpool
Simpson, A. Hope, Annandale, Sefton Park
Willmer, Miss J. H., 20, Lorne-road, Oxton, Birkenhead

C. Honorary Members.

H.S.H. Albert I., Prince of Monaco, 25, Faubourg St.
    Honore, Paris
Bornet, Dr. Edouard, Quai de la Tournelle 27, Paris
Claus, Prof. Carl, University, Vienna
Fritsch, Prof. Anton, Museum, Prague, Bohemia
Giard, Prof. Alfred, Sorbonne, Paris
Haeckel, Prof. Dr. E., University, Jena
Hanitsch, R., Ph.D., Raffles Museum, Singapore
Leicester, Alfred, Buckhurst Farm, nr. Edenbridge, Kent
Solms-Laubach, Prof. Dr., Botan. Instit., Strassburg

REPORT of the LIBRARIAN.

Although no additional exchanges of publications have been arranged since the last Annual Report, the number of Societies and Institutions exchanging with us is still maintained and is the same as that given last year, viz. :—seventy-six.

The following list gives the titles of the exchanges and donations received during the session :—

1. Allgemeine Fischerei-Zeitung. Vol. XX., Nos. 10—26, Vol. XXI. Nos. 1—5.
2. Amsterdam, Verhand. der K. Adad. van Wetenschappen (Ser. 2). Vol. IV., Nos. 1—6 ; Jaarboek, 1894.
3. Annaes de Sciencias Naturaes. Vol. II., Nos. 3 and 4 ; Vol. III., No. 1.
4. Archives Neerlandaises de Sciences exactes et naturelles Vol. XXIX. Nos. 2 and 5.
5. Australian Museum, The ; Report of the Trustees for the year 1894.
6. Bergen, Museums Aarbog, 1894-5.
7. Berlin, Math. u. natur. Mittheilungen d. k. preuss. Akademie, Nos. 3—7.
8. Berlin, Sitzungsberichte d. k. preuss. Akademie, Nos. 1—38.
9. Bihang till K. Svenska Vetenskaps—Akad. Handl. Vol. XX. Zool. pt. IV., Bot. pt. III.
10. Bordeaux, Procès—Verbaux de la Société Linnéenne. Vol. XLVII., 1894.
11. Bulletin Scientifique de la France et de la Belgique. Vol. XXVII., part 1.
12. Canada, Proceed. and Trans. of the Nova Scotian Institute of Science. Vol. VIII., pt. 4.
13. Canada, Bulletin Natural History Society of New Brunswick. No. 13.
14. Chili, Actes de la Société Scientifique. Vol. IV., No. 5.
15. Chili, Revista Chilena de Hijiene. Vol. I., No. 3.
16. Denmark, Oversigt over det K. Danske Vidensk. Selskabs. Forhand. 1894, No. 3 ; 1895, Nos. 1—4.
17. Denmark, Oversigt over det K. Danske Vidensk. Memoires. Ser. 6. Vol. VIII., No. 1.

18. Frankfurt, Senckenbergische Naturforschende Gesellschaft. 1895.

19. Gottingen, Nachrichten d. k. Gesellschaft d. Wissenschaften. 1895, Nos. 1—4.

20. Japan, Journal of the College of Science, Imperial University. Vol. VII., parts 4 and 5.

21. Kjobenhavn, Report of the Danish Biological Station to the Home Department V. 1894, Vol. VIII., 2 ; Vol. IX., 1.

22. Leipzig, Berichte d. k. Gesellschaft d. Wissenschaften. 1894, No. 3 ; 1895, Nos. 1—4.

23. Manchester, Microscopical Society, Transactions and Annual Report, 1894.

24 Marine Biological Association of the United Kingdom, Journal, N.S. Vol. III., Vol. IV., Nos. 1 and 2.

25. Moscou, Bulletin de la Société Impériale des Naturalistes. 1895, Nos. 1 and 3.

26. Museum d'Histoire naturelle, Paris, Bulletin. 1895, Nos. 1, 3, 6 and 8.

27. Nancy, Bulletin de la Société des Sciences. (Ser. 2), Vol. XIII., pt. 29, Nos. 4—6 ; 8—10 ; Vol. II., part 1.

28. Napoli, Rendiconto dell Accademia dell Scienze fisiche e matematiche. (Ser. 3), Vol. 1.

29. Natuurkundig Tijschrift voor Nederlandsch. Indie. Vol. LIV ; Vol. III. (N.S.).

30. New Zealand Institute, Trans. and Proceedings. Vol. XXVII., 1894.

31. Rheinlande, Verhand. d. naturh. Vereins d. preuss. Vol. LII., part 1.

32. Royal Microscopical Society, Journal. 1895, parts 4—6 ; 1896, part 1.

33. Scottish Fishery Board, Annual Report. XIII.

34. Stavanger Museum. Aarsberetning for 1894.

35. St. Pétersbourg, Bulletin de l'Accadémie Impériale des Sciences. (Ser. 5), Vol. II., Nos. 4 and 5 ; Vol. III., No. 1.

36. The Naturalist. May to December, 1895. January to April, 1896.

37. Torino, Bollet. dei Musei di Zoologia ed Anatomia comparata delle R. Universita. Vol. X., Nos. 193—209.

38. United States, Bulletin of the Fish Commission. Vol. XIV., 1894.

39. United States, Proceedings of the National Museum. No. 1018—1032.

40. United States, Bulletin of the National Museum. No. 39, parts J. and K., No. 48.

41. United States, Report of the National Museum. 1893, pp. 337—487 ; 587—624 ; 653—663.

42. United States, Annual Report Museum of Comparative Anatomy. 1894-95.

43. United States, Bulletin Museum of Comparative Anatomy. XXV., No. 12; XXVI., Nos. 2 and 15; XXVII., Nos. 1—6; XXVIII., No. 1.

44. United States, Proceedings of the Academy of Sciences, Philadelphia. Vol. III., parts 1 and 2.

45. United States, Proceedings of the Boston Society of Natural History. Vol. XXVI.

46. United States, Transactions of the Academy of Science, St. Louis. Vols. II.—V.; Vol. VI., No. 18; Vol. VII., Nos. 1—3.

47. United States, Tufts College, Mass. Studies No. IV.

48. Upsala, Royal Society of Sciences (Royal University). Ser. III., Vol. XV., No. 2.

49. Upsala, Studien über Nordische Actinien, 1 von Oskar Carlgren.

50. Upsala, Studier ofver Alkaloidernas Lokalisation,—Familjen Log-aniaceae. By Marten Elfstarnd.

51. Upsala, Anatom. System. Studier ofver Lokstammiga Oxalisarter. By Th. Fredrikson.

52. Upsala, Den Gotländska vegetationens ütavecklingshistoria. By Rutger Sernander.

53. Upsala, Carl von Linné, II. By Th. M. Fries.

54. Upsala, Naturalhistorien 1 Sverige. By Th. M. Fries.

55. Upsala, Uber der Rhizoidenbildung. By O. Borge.

56. Upsala, Uber *Echinorhynchus turbinella*, *E. brevicollis*, und *E. porrigens*. By Ernst Borgström.

57. Victoria, Proceedings of the Royal Society. Vol. VII.

58. Wien, Verhandl. d. k. Zool-botanischen Gesellschaft. Vol. XLV., parts 4 and 5, 8—10; Vol. XLVI., part 1.

59. Wien, Annal. d. k. k. Naturh. Hofmuseums. Vol. IX., parts 1—4.

60. Zurich, Vierteljahrschrift d. naturf. Gesellschaft. Vol. XL., No. 2.

61. Zeitschrift für fischerei. 1895, part 6.

By Donation.

62. The Irish Naturalist, Special Galway Conference Number. Vol. IV., No. 9., 1895.

63. Le Laboratoire Maritime du Muséum de Paris. Par A. E. Malard.

64. Brit. Assoc. for the Advancement of Science, Ipswich 1895, Address to the Zool. Sect. By William A. Herdman, D.Sc., F.R.S., F.L.S., F.R.S.E., Professor of Natural History in University College, Liverpool, President of the Section.

65. The Marine Zoology, Botany, and Geology of the Irish Sea, Third Report of the Committee.

66. On Oysters and Typhoid; an experimental enquiry into the effect upon

the Oyster of various external conditions including Pathogenic organisms. By Professors Boyce and Herdman.

67. Some facts and reflections drawn from a study in Budding in Compound Ascidians. By Professor W. E. Ritter, University of California.

68. Occupation of a table at the Zoological Station, Naples. Report of the Committee.

69. Investigations made at the Laboratory of the Marine Biological Association at Plymouth. Report of the Committee.

70. Fauna of Liverpool Bay. Report IV.

71. Science, N.S. Vol. II., No. 41.

Nos. 62—71. Presented by.Professor Herdman, F.R.S.

72. Sur un examplaire Chilien de *Pterodela pedicularia*, L., à nervation doublement anormale. Par Alfred Giard, presented by the Author.

73. Etudes sur les Fourmis les Guêpes et les Abeilles 8e Note.

74. Sur l'organe de nettoyage tibio tarsien de *Myrmica rubra*, L., race *Levinodis*. Nyl.

75. Neuvième note—sur *Vespa crabro*, L., Histoire d'un nid depuis son origine.

76. Onzieme note sur *Vespa germanica* et *V. vulgaris*.

77. Dixiéme note sur *Vespa media*, *V. silvestris* et *V. saxonica*.

78. Sur la *Vespa crabro*, L. Ponte, conservation de la chaleur dans la nid.

79. Sur les nids de la *Vespa crabro*, L., ordre d'apparition des alveoles.

80. Observations sur les Frelons.

Nos. 73—80. Par Charles Janet and presented by the Author.

81. Flora of Anglesey. Griffith.

82. The Amphipoda of Bate and Westwood's "British Sessile-eyed Crustacea." By Alfred O. Walker.

Nos. 81—82. Presented by A. O. Walker, Esq., F.L.S.

83. The Recent and Fossil Flora and Fauna of the Country around Liverpool. By G. H. Morton, .F.G.S., presented by the Author.

84. Résultats des Campagnes Scientifiques accomplies sur son yacht. Par Albert 1ᵉʳ Souverain de Monaco. Fasc. VIII. Zonthaires provenant des campagnes du yacht l'Hirondelle. Par E. Jourdan.

Fasc. I. Contribution l'etude des Céphalopoda de l'atlantique Nord. Par Louis Jouhin.

Presented by H.S.H. the Prince of Monaco.

85. Some recent evidence in favour of Impact. By A. W. Bickerton.

86. Copy of letters sent to "Nature" on Partial Impact. By Professor Bickerton.

List of Societies, etc., with which publications are exchanged.

AMSTERDAM.—Koninlijke Akadamie van Wettenschappen.

Koninklijke Zoölogisch Genootschap Natura Artis Magistra.

BALTIMORE.—Johns Hopkins University.

BATAVIA.—Koninklijke Natuurkundig Vereeniging in Ned. Indie.

BERGEN.—Museum.

BERLIN.—Konigl. Akadémie der Wissenschaften.

Deutscher Fischerei-Vereins.

BIRMINGHAM.—Philosophical Society.

BOLONGA.—Accademia delle Scienze.

BONN.—Naturhistorischer verein des Preussichen Rheinlande und West-falens.

BORDEAUX.—Société Linnéenne.

BOSTON.—Society of Natural History.

BRUSSELLS.—Académie Royal des Sciences, etc., et Belgique.

CAMBRIDGE.—Morphological Laboratories.

CAMBRIDGE, Mass.—Museum of Comparative Zoology of Harvard College.

CHRISTIANIA.—Videnskabs-Selskabet.

DUBLIN.—Royal Dublin Society.

EDINBURGH—Royal Society.

Royal Physical Society.

Royal College of Physicians.

Fishery Board for Scotland.

FRANKFURT.—Senckenbergische Naturforschende Gesellschaft.

FREIBURG.—Naturforschende Gesellschaft.

GENEVE.—Société de Physique et d'Histoire Naturelle.

GIESSEN.—Oberhessische Gesellschaft für Natur nnd Heilkunde.

GLASGOW.—Natural History Society.

GOTTINGEN.—Konigl. Gesellschaft der Wissenschaften.

HALIFAX.—Nova Scotian Institute of Natural Science.

HARLEM.—Musée Teyler.

Société Hollandaise des Sciences.

KIEL.—Naturwissenschaftlichen vereins fur Schleswig—Holstein.

KJOBENHAVN.—Naturhistorike Forening.

Danish Biological Station, (C. G. John Petersen).

Kongelige Danske Videnskabernes Selskab.

LEEDS.—Yorkshire Naturalists Union.

LEIPZIG.—Konigl. Sachs. Gesellschaft der Wissenschaften.

LILLE.—Revue Biologique du Nord de la France.

LONDON.—Royal Microscopical Society.

British Museum (Natural History Department).

MANCHESTER.—Microscopical Society.

        Owens College.

MARSEILLE.—Station Zoologique d'Endoume.

        Musée d'Histoire Naturelle.

MASSACHUSETTS.—Tufts College Library.

MECKLENBURG.—Vereins der Freunde der Naturgeschichte.

MELBOURNE.—Royal Society of Victoria.

MONTPELLIER.—Académie des Sciences et Lettres.

MOSCOU.—Société Impériale des Naturalistes.

NANCY.—Sociéte des Sciences.

NAPOLI.—Accademia delle Scienze Fisiche e Matematiche.

NEW BRUNSWICK.—Natural History Society.

OPORTO.—Annaes de Sciencias Naturaes.

PARIS.—Museum d'Histoire Naturelle.

      Société Zoologique de France.

      Bulletin Scientifique de la France et de la Belgique.

PHILADELPHIA.—Academy of Natural Sciences.

PLYMOUTH.—Marine Biological Association.

ST. LOUIS, MISS.—Academy of Science.

ST. PETERSBURG.—Académie Impériale des Sciences.

SANTIAGO.—Société Scientifiq du Chili.

STAVANGER.—Stavanger Museum.

STOCKHOLM.—Académie Royale des Sciences.

SYDNEY.—Australian Museum.

TOKIO.—Imperial University.

TORINO.—Musei de Zoologia ed Anatomia Comparata della R. Universita.

TORONTO.—Canadian Institute.

TRIESTE.—Societa Adriatica de Sceinze Naturali.

UPSALA.—Upsala Universitiet.

WASHINGTON.—Smithsonian Institution.

        United States National Museum.

        United States Commission of Fish and Fisheries.

WELLINGTON, N.Z.—New Zealand Institute.

WIEN.—K. K. Naturhistorischen Hofmuseums.

     K. K. Zoologisch—Botanischen Gesellschaft.

ZURICH.—Zürcher Naturforschende Gesellschaft.

Dr.   In Account with ISAAC C. THOMPSON, Hon. Treasurer.   **Cr.**

1896.	£	s.	d.
To Use of Rooms, University College	3	3	0
„ Tea and Attendances at Meetings	3	11	0
„ Printing and Stationery—			
Dobb, T. & Co.	62	19	0
Brown, S. J.	4	16	0
Walmsley, G. G.	0	8	0
„ Library Book Case	5	0	0
„ Postage and Carriage of Vols., &c, Hon. Librarian	6	9	10
„ „ Hon. Sec.	3	9	4
„ „ Hon. Treas.	0	15	0
„ Balance in hand, September 30th, 1896............	55	13	4
	£146	4	6

1896.	£	s.	d.
By Balance, 30th September, 1895............	65	7	2
„ 7 Members' Entrance Fees at 10/6............	3	13	6
„ 2 Student Members' Fees at 2/6............	0	5	0
„ 64 Members' Subscriptions at 21/-............	67	4	0
„ 13 Student Members' Subscriptions at 5/-............	3	5	0
„ Sale of Proceedings............	6	9	10
	£146	4	6

By Balance in hands of Treasurer............ ...   £55 13 4

ISAAC C. THOMPSON,
Hon. Treasurer.

Audited and found correct,
A. T. SMITH, Junr.

LIVERPOOL, *September 30th, 1896.*

TRANSACTIONS

OF THE

LIVERPOOL BIOLOGICAL SOCIETY.

PRESIDENTIAL ADDRESS

ON

BOTANIC GARDENS—PAST AND PRESENT.

BY

R. J. HARVEY GIBSON, M.A., F.L.S.,

PROFESSOR OF BOTANY IN UNIVERSITY COLLEGE, LIVERPOOL.

MY first duty is to thank you formally but none the less sincerely, for the honour you have done me in electing me to fill the presidential chair of the Liverpool Biological Society. I thank you not only for the honour you have done me as an individual but for your implied recognition of the importance of the science which I represent in this College. I need hardly say that it shall be my constant endeavour to fulfil to your satisfaction the duties devolving on your President, and I confidently look for your interest and cooperation in aiding me to maintain the high standard of work which this Society has always aimed at under the guidance of the distinguished biologists who have preceded me in the occupancy of this chair.

Our meetings, I take it, have two important functions to fulfil, the one to awaken in others and to stimulate in ourselves interest in the science of Biology, the other to advance that science by original investigation and research. Whilst frankly acknowledging the necessity for endeavouring to render our meetings as generally attractive and interesting as possible, I believe that we would be taking a distinctly retrogressive step were we to permit any such aim to obscure the wider and more important purpose of our Society, viz., to attempt, however feebly, still conscientiously, to elucidate some problem or add some fact to our knowledge of living organisms, by original contributions to biological literature.

Probably the greatest scientific work of the incoming century will be done not so much within the realm of one particular science as on the borderland where two or more sciences meet. That research is steadily tending in this direction is becoming more evident day by day. Recognising this trend in scientific research, we as a society may, it seems to me, considerably expand our own stock of conceptions and ideas by occasional joint discussions with the devotees of allied sciences on subjects of mutual interest. An initiative was taken in this direction last session when the Physical Society united with us in a consideration of certain problems connected with the physiology of the organ of hearing. I trust it may be possible during the present session to arrange a joint meeting with some other scientific society whose sphere of work to some extent may overlap that in which we biologists labour.

Again, I believe that it is beneficial to the welfare of our Society that we should occasionally go outside our own circle of members and invite some distinguished biologist to address us on a subject which he has made specially his own. On several occasions in past years we have had the pleasure of listening to such addresses, and the meetings at which we have been so favoured have been amongst the most successful of any we have held. I am hopeful that the precedent so established may be again acted upon during the year of my presidency.

I would repeat, however, that in my opinion neither of these methods of quickening public interest in Biology must be allowed to obscure our main object and aim, viz., the production of original research and the publication of our yearly volume of Transactions, which I am proud to to think have already won the approbation of biologists both in Europe and America.

It is fortunate for me that your President is permitted some latitude in his choice of a subject on which to address you, and some freedom in his method of treatment of it. I could scarcely hope to interest you in the detailed anatomical researches to which I have devoted what little time I can spare from professorial duties; on any other subject I must perforce speak at second hand. I would beg your forbearance, therefore, if I venture to offer you a necessarily brief historical and descriptive account of some of the great botanical gardens of the world, and endeavour to shew you how the idea of a botanic garden first arose in the minds of the medieval botanists and thereafter became concrete in the continent of Europe and in our own country. I trust that some, at least, of the remarks I have to make may be new to you; and if I fail to make the subject interesting you will I hope lay the blame on the teller not on the story.

It requires no great powers of observation and of reasoning to enable us to classify gardens as either useful, ornamental or scientific. Not that a scientific garden is not or cannot be ornamental and even economically useful, or a useful garden at once scientific and ornamental, but simply that in planning out a garden we are accustomed to consider primarily whether it is to aim at being a decorative appendage to a house or a city, an area set a-part for the cultivation of pot herbs or a living museum for study and research. Scientific or Botanic Gardens proper have for their chief purpose the advancement of the science of Botany, are therefore just as old as the science itself and naturally have developed *pari passu* with it. An outline knowledge of the history of Botany will obviously aid us greatly in understanding the developmental stages through which the botanic garden has passed.

It would be quite beside my purpose to refer even in brief to the botanical writings of the Ancients. The names of Aristotle, Theophrastus and Dioscorides are perhaps those most familiarly associated with the early history of Botany, although the classical scholar is not unacquainted with the contributions to knowledge of plant lore of such writers as Empedocles, Anaxagoras, Apollodorus, Democritus and the numerous writers on agricultural topics down to the times of Vergil and Pliny. A gap of several centuries follows destitute of a single botanical publication of note until we reach the 14th and 15th centuries when we meet with the records left us by the early herbalists. Even at the best a herbal of that time was little more than a catalogue of plants used in pharmacy—a list of simplicia from which the first dispensers compounded their drugs. Indeed as Sachs puts it " the chief object of the earlier herbalists was to rediscover the plants employed in medicine by the physicians of antiquity, and, if possible, to identify in the west the plants mentioned as of medicinal value by the Greeks."

The later herbalists aimed at something higher. They rightly relinquished the study of classic texts for the study of nature, and contented themselves with recording as accurately as possible the external configuration of the plants growing in or near the districts in which they resided. It is true again that their labours resulted in a series of descriptive and illustrated catalogues with little or no pretence at systematic arrangement of subject matter. They confined themselves to diagnosing and illustrating individual form and to enumerating real or supposed medicinal virtues.

From a knowledge of the history of science in general

one learns to expect that a long patient and apparently purposeless accumulation of data often finds at last its justification and its fulfilment in the sudden and brilliant discovery of one or more grand natural laws. It is somewhat remarkable therefore that the idea of natural affinity arose only incidentally in the process of cataloguing and describing plants and was, even by those who first felt it, looked upon as of quite secondary importance. Almost, one might say without conscious intention, but yet, as we now see, by a very natural association of ideas, the writers of the later herbals, for instance Bock, Fuchs, and Brunfels, described successively plants which had an obvious family likeness to each other. A most important step was thus gained and one which led quite easily and naturally to the deliberate subordination of the question of pharmaceutical utility and to the elevation to first rank of the study of plant relationship and of the morphological features on which that principle was based. To Kaspar Bauhin belongs the credit of first clearly recognising this principle and of emphasising it in his works. To this botanist and not to Linnaeus we are indebted for the first appreciation of the value of generic and specific names: he it was who first made use of the binary system of nomenclature in describing living objects. Grotesque as his classification may seem to students of these later days, still in Bauhin's *Pinax* and in the *Prodromus Theatri Botanici* we meet for the first time an attempt at grouping plants in accordance with natural affinity, not merely at cataloguing in alphabetical order. Bauhin was born at Bâle in 1550 and died in 1624, and the period of his life may be said to include the date of the nativity of Botany as a Science.

Another great name of the 16th Century that stands out preeminently is that of Andraea Caesalpino, Professor of Botany in the University of Pisa. Not content with mere analysis and observation, Caesalpino brought to the study of plants an intimate acquaintance with Aristotelian philosophy, and his works teem with theoretical discussions on the seat of the soul in plants, the doctrine of metamorphosis and such like products of scholasticism. His classification is based not upon deduction from observed natural phenomena but on philosophic abstraction and reasoning on the nature of plants, and the relative value of their various parts. In this respect Caesalpino sounded a theme on which his successors played variations for more than a hundred years until Linnaeus brought the concerto to a finale in the *Systema Naturæ*. In justice to Linnaeus one must bear in mind that he himself proposed his artificial system avowedly as provisional; that that system should have been obstinately not to say pugnaciously adhered to by his followers for well nigh a century after its illustrious author had passed away, and that his own suggestive speculations as to the true principles of classification, given to the world in the *Philosophia Botanica*, should have been almostly completely ignored, is due to the blind reverence with which his disciples regarded one who had accomplished what we must recognise as a colossal task.

Long before the second revival of Botany, which took place at the beginning of the 18th century, travellers like Clusius and Albini had lifted the veil from a new world of plant life, and anatomists like Malpighi and Grew, with their new found microscopes were endeavouring to unravel the mysterious histological complex that these instruments revealed not only in the phanerogamic world but in the hitherto untrodden field of cryptogamic Botany, wherein Micheli and Dillenius stand out prominently as pioneers. The last days of the 17th and early days of the 18th century were noteworthy for yet another development—the birth of vegetable physiology which dates from Kölreuter and Sprengel's observations and experiments on fertilisation and Stephen Hale's classic labours on the movements of sap. The De Jussieus and later the De Candolles inaugurated a new epoch in systematic Botany by the publication of their natural systems of taxonomy, based on the essential principle that natural affinities are not to be determined by any one character but by the sum of all, whilst the developmental and physiological methods were further emphasised by the publication in 1842 of Schleiden's *Principles of Botany* wherein the modern standpoint is at last clearly enunciated and defined. With the acceptance by botanists of these principles we reach comparatively recent times, for the next great names that meet us are those so familiar to the botanical student of to-day.

Botany then, as an organised study of the structure, functions and mutual relationships of plants had Medicine and Pharmacy for its parents, although I fear its parents are nowadays rather inclined to ignore their offspring. The search for simples amongst the herbs of the field resulted in the discovery of a great Science; the need for the study of those simples brought about the foundation of gardens to which the ancient apothecary might have ready access and from which he might conveniently obtain the particular drug he required. The finds of travellers involved successive extensions in the superficial area of the gardens, and the birth of the new interest in plants for their own sake revolutionised their arrangement. Microscopic and experimental research demanded laboratories in which the student might explore the minute structure of plants or attempt to solve the thousand and one problems in vegetable physiology, whilst the growth of popular interests and the necessities of international commerce encouraged the foundation of museums of Economic Botany. In short all the progressive phases in the development of the Science of Botany had their concrete counterparts in the evolution of the Physic Garden of the 15th century into the scientific Botanic Garden of the 19th.

In the time that is placed at my disposal this evening I desire to give you some notion of this progressive development and to describe to you a few illustrative examples of the Physic Gardens of the Herbal period, of the Botanic Gardens of the 17th and 18th centuries and finally of the modern Botanic Gardens both in our own country and abroad.

It is not necessary for my present purpose to do more than name some of the gardens of classical times. I do not refer of course to the numberless private gardens attached to the palaces and villas of the Greek and Roman nobles, but to the public gardens, records of which have come down to us in the classics. Amongst these perhaps the most famous are the gardens of Theophrastus at Athens, and of Ptolemy Philadelphus at Alexandria, of King Attalus at Pergamos, and the gardens of Carthage. We read also of a medicinal garden belonging to Antonius Castor at Rome, perhaps the first of its kind, and Vergil, Martial and other Latin writers frequently refer to gardens devoted to the culture of the vegetable delicacies most fancied by the palates of the Roman epicures.

In the early middle ages the various monkish orders were, as every one knows, the custodians of the degenerate knowledge and practice of medicine transmitted from the later Latin authors, until the quaint mixture of ancient science and medieval black art was at last superseded by the Benedictines, whose academy at Monte Cassino in

Campania, was the seat of the revival of the art of healing founded by Hippocrates and Galen. The *Civitas Hippocratica* as it was called, was a non-religious establishment, where law and philosophy, but especially medicine, were taught, and to which came students from all parts of the then known world. In connection with this Institution we find records of a Physic Garden founded in 1309 by a certain Matthaeus Sylvaticus and in 1333 we read of another Medicinal Garden established by the famous Venetian Doctor Gualterus on an area given him by the Republic " pro faciendo hortum pro herbis necessariis artis suæ."

Data are more abundant for dealing with the history of the University Gardens of the 16th century and to these we may now turn.

Whilst Fuchs and Brunfels were endeavouring to weed out the rank growths that had been assiduously cherished by the earlier herbalists, Francesco Buonafede, the first professor of Botany in the first university in Europe, that that of Padua, obtained from the Senate of Venice in 1545 a grant of land for the establishment of a medicinal garden. That garden exists to the present day, and occupies the same (though now greatly extended) site granted in the 16th century by the enlightened Signiory.

The foundation of the physic garden of Padua was the signal for the establishment of gardens in several other Italian cities, and in quick succession we read of gardens laid out at Pisa in 1547, at Florence in 1550, at Rome in 1566, and at Bologna in 1567. It is impossible for me in the time at my disposal to describe to you even briefly the history of these and of many other Italian gardens founded at a later date. It must suffice merely to note that the example of the Italian States was soon followed in Northern and Western Europe, and before the end of the century the great gardens of Leyden and Montpellier had been founded.

Just as the travels of Lusitanus, Calceolari, Ghini, Aldrovandi and others materially aided the extention and enlargement of the gardens of Padua, Pisa and Bologna, so in the north, through the efforts of Dodonaeus, Lobelius and Clusius a medicinal garden was in 1577 founded in connection with the famous university of Leyden at the expense of the Municipal Council of the city.

If Padua, Pisa and Bologna were the great botanical schools of the east, as Leyden was that of the north, Montpellier claimed the premier place in the west. The names of the renowned botanists Fuchs, Gesner, Clusius, Lobelius and Bauhin appear on its list of students although it did not succeed in retaining them amongst its teachers. The new and already famous gardens of the northern and eastern universities drew away the cream of the botanical students from its halls. Montpellier could not afford to be behind hand : it must needs have a botanical garden also, and so there came into existence in 1596 a garden destined to eclipse even the most famous of its predecessors in Holland and Italy. For this reason I have selected it for fuller description. Its founder was Pierre Richer de Belleval, a native of Picardy and graduate of Montpellier University. At his instigation the Parliament of Languedoc granted permission and land to found a garden in connection with the medical faculty of the University. Belleval's chief aim, as he himself states, was to cultivate plants both native and exotic, under conditions similar to those of nature. He says, "J'ai exécuté vos ordres et fondé sous le nom de Jardin royal, un établissement digne d'un grand empire. Il est divisé en plusieurs parties présentant chacune une exposition différente ; un monticule offre deux versants tournés l'un vers le Sud,

l'autre vers le Nord. Il y a dés lieux escarpés, des rochers,
des sables exposés au soleil, d'autres ombragés, humides,
inondés, ou d'un sol fertile ; on y trouve des buissons, des
mares, des marais, dans lesquels les végetaux herbacés, les
sous-arbrisseaux et les arbres prospèrent admirablement."
(*Onomatologia*, Pref. 1598).

The main gateway of the garden led into an avenue of
laurels, to the north of which lay the medical garden with
the following inscription over the entrance : PLANTÆ
QUARUM IN MEDICINA HIS TEMPORIBUS MAXIME USUS EST.
The pharmaceutical collections were arranged in rows in
alphabetical order. North of the medicinal garden came
a six-tiered hill. The lowest tiers facing the south were
occupied by PLANTÆ ODORATÆ VENENATÆ UMBELLIFERÆ
SPINOSÆ CATHARTICÆ SCANDENTES ALIIS INNIXÆ ; the
north side of the mound by PLANTÆ QUÆ IN ASPERIS
SAXOSIS APRICIS ET IN IPSO LITTORE NASCUNTUR, whilst
on the top were grown PLANTÆ QUÆ IN CLIVIS MONTIBUS
FRUTETIS DUMETIS ET SABULOSIS ADOLESCUNT. Over
the chief gate was engraved the arms of France and the
name LE JARDIN DU ROY, 1596 ; whilst underneath was
a rather more poetic and classic rendering of our "visitors
are requested not to touch"—HIC ARGUS ESTO NON
BRIAREUS. The plate, I ought to add, represents only
part of the garden. In describing it Belleval says, "Le
Jardin du Roi était coupé en dieux parties ; l'une appelée
le Jardin médical, l'autre la pépinière ; la première était
destinée aux démonstrations des plantes et consacrée à
l'Université ; l'autre était rempie des plantes étrangères,
qui, pour la plupart, étaient montagneuses, destinées
plutôt à la curiosité qu'a la nécessité, afin que ceux qui
accouraient des provinces et des nations étrangères, y
reconnussent leurs richesses." A herbarium and audi-
torium, with a dedication to Henry of Navarre, occupied
the extreme north eastern corner of the garden.

The gardens of Montpellier created a great sensation among the botanists of the time, all of whom speak of it in terms of the highest admiration. Despite lack of funds Belleval added rapidly to the contents of his garden, devoting special attention to the collection of local plants. The siege of Montpellier in 1622 inflicted great damage on the garden, but the courageous founder did not lose heart. No sooner had he seen the last of the troops than he set to work again and in his labours was successful in interesting the great Richelieu. He had scarcely completed the renovations necessary when he died in 1632. It will be unnecessary for me to describe in detail the gradual progress of the garden during the century following the death of Belleval. The arrangement of the plants had been twice changed, once to conform to the system of Tournefort, and again to bring it into conformity with the artificial system of Linnaeus. The beginning of the 19th Century saw the garden under the direction of the famous A. P. de Candolle, and his election to the directorate was followed by an entire rearrangement on the principles of the natural system of classification now so well known and universally followed. De Candolle added an arboretum and greenhouses and in many other ways extended the scientific usefulness of the gardens. His reign at Montpellier lasted only 6 years, for in 1816 he was appointed to the chair of Botany in Geneva, where he spent the remainder of his life.

A word finally as to the present condition of the garden. It includes about 45,000 sq. metres divided into three regions; first, to the south, a triangular region devoted to the uses of the School of Medicine, a median oblong (the mount of Belleval's time), and a northern area irregular in outline added to the garden as an arboretum in De Candolles time. Between the systematic garden and the mound (which bears a miscellaneous collection of trees and shrubs) is the orangery and a line of greenhouses for temperate plants. The northern area includes an acclimatisation garden and an area devoted to the culture of fruit trees and plants used in the arts. The extreme north is occupied by the necessary store houses and out-buildings, the museum, library, &c.

The decade which saw the foundation of the Montpellier gardens saw also the establishment of gardens in connection with the Universities of Heidelberg, Leipzig and Breslau. In 1597 Paris followed suit. The story goes that the now famous Jardin des Plantes was primarily established with the intention of investigating what variations were possible in the style of the bouquets worn at the Royal Court—truly a noble piece of research! Its first claim to be considered as a botanic garden was not made until about 1630 when professorships of Botany and Pharmacology were instituted. Strassburg in 1619, Jena in 1629, Oxford in 1632, Messina in 1638, Chelsea in 1677, Edinburgh in 1680 and Berlin in 1714 are the next examples of note. As the Physic Garden at Chelsea has the honour of being the first and oldest botanic garden in Great Britain, perhaps I may select it for more than mere mention. It is true that Gerard, the author of the famous Herbal, is said to have had a private medicinal garden attached to his house in Holborn, and also Tradescant, gardener to James I., and after whom the well known *Tradescantia* has been named, cultivated exotics in a garden situated where S. Lambeth now stands, still as a public garden that of Chelsea is undoubtedly entitled to priority. The immediate founders of the Chelsa Gardens were the members of the Society of Apothecaries who now for over two centuries have maintained it solely for the advancement of the science of Botany. The name " Physic Garden " by which it was for long known was to a certain extent a misnomer. Medicinal plants were certainly cultivated in it but in the deed of gift by Sir Hans Sloane the growing of plants for pharmaceutical trade purposes is expressly forbidden.

The history of the first beginings of the Chelsea Gardens sounds rather paltry not to say ridiculous after one has studied the story of the foundation of the gardens of Padua and Montpellier. We read that the ground on which the Chelsea Garden is now laid out was originally taken by the Apothecaries Society as a spot on which to build a convenient barge house for the ornamental barge which the society (like other city companies) then possessed. This plot of ground was walled round in 1674, and four years later appears to have been planted with fruit trees and herbs for the use of the Apothecaries' laboratory. A greenhouse and stove were added shortly afterwards and the Society then began to exchange plants with the gardens of the University of Leyden. The purchase of the estate by Dr. (afterwards Sir Hans) Sloane in 1712 at once brought about a change for the better in the prospects of the Chelsea establishment. Sloane was a scientist of no mean rank—a pupil of the chemist Stahl and of the botanist Tournefort and further a friend of Ray, the father of English Botany. His scientific achievements won for him the presidency of the Royal Society in succession to Sir Isaac Newton. His enthusiastic love for science in general and for Botany in particular led him to readily accede to the request of the Apothecaries Society for liberal treatment with regard to the conditions of tenure of the garden now included within his estate, and in 1722 Sloane generously handed over to them the ground—an area of over three acres—on condition, however, that the Society paid an annual rent of £5, and rendered yearly to the Royal Society 50 species of distinct plants, well dried, preserved and named, which had been grown in the garden that same year—those presented in each year to be specifically different from those of every former year, until 2000 had been delivered.

One gets a fairly good idea of the contents of the Chelsea Garden at this time from the Catalogue published in 1730 by Miller, the then chief gardener. The plants were classified into herbs and under shrubs, shrubs and trees each section being arranged alphabetically. Four hundred and ninety nine plants in all are named and are chiefly those which were employed in the pharmaceutical preparations of that day. Hothouses for the cultivation of exotics were opened in 1732. Four years later Linnaeus visited the garden, and no doubt his interest in it did much towards hastening the introduction of the Linnaean method into the systematic arrangement of the plants in the gardens. Amongst the directors of Chelsea Garden appear the names of some of the foremost of our English botanists—Sherard, whose generosity established the Sherardian professorship at Oxford—Hudson, the author of the *Flora Anglica*—Curtis, the editor of the familiar *Botanical Magazine*, and Lindley, whose contributions to systematic Botany and to Taxonomy are too well known to need enumeration.

The gardens are now more of historic than of scientific importance, being dwarfed into insignificance by the great, but comparatively modern, establishment at Kew. It now covers about four acres along the Chelsea embankment. The centre is occupied by the statue of Sir Hans Sloane; the upper part of the garden contains the collection of medicinal plants, whilst towards the river are arranged beds of hardy herbaceous plants in natural orders, with trees and shrubs interspersed, the remains of former plantings. The houses occupy the north west end.

A few years before the foundation of the Chelsea Garden
that more especially associated with the name of Linnaeus
came into being. Considering the enormous influence
this famous man had on Botany, a brief account of the
garden at Upsala may not be without interest. Although
founded by Rudbeck in 1660 it was not until 1742 that
the Upsala garden came under the direction of Linnaeus.
An old pamphlet dated 1745 published by his authority
gives an account of the gardens at that date. You will
be able to obtain some notion of their character from the
photo of the plate which illustrates this work. The
garden was in the form of an oblong. The northern end
was lined with greenhouses—cool, temperate and hot.
To the right of this range was an "area vernalis" to the
left an "area autumnalis" and immediately in front an
"area meridionalis."   In the middle of the garden were
three aquaria for marsh, lake and river plants respectively.
The remainder of the space was occupied by two great
sets of beds, that to the right for annuals, that to the left
for perennials, whilst to the south were further ornamental
flower beds, the houses of the gardeners, the museum,
library and so on.

It is worth pausing to note at this point that whilst, as we
have already seen, the first Italian Gardens were avowedly
purely medicinal, that of Montpellier was a compromise.
Although laid out primarily for the cultivation of medicinal
herbs, it was not purely and exclusively pharmaceutical.
In accordance with Belleval's scheme a portion of the
space alloted to him was designed to accommodate plants
not necessarily medicinally useful and arranged according
to their habit, rock, aquatic, marsh loving, and so on as
the case might be.   Here in the Upsala garden the
specially pharmaceutical purpose is ignored and the beds

are laid out entirely to suit the botanical character of the plant, and in accordance with its natural habit and in no respect with its usefulness to man. These three historic gardens Padua, Montpellier and Upsala, may indeed be taken as illustrating three great stages in the evolution not only of the botanic garden itself but also of the science of Botany. Established first of all in Italy as a living herbal, as a concrete illustration of Medicinal Botany, the garden reflected in itself the birth of the science when it became planned, as at Montpellier, partly on utilitarian, partly on physiological lines, and becomes finally in Upsala a purely educational establishment conceived and carried out in accordance with scientific principles.

To attempt to follow the history or even enumerate the botanical gardens founded in other cities in the end of the last and the beginning of this century, and to trace how the hard and formal scientific conception of a botanic garden became artistically modified and softened, would lead me far beyond the limits of time which your courtesy grants me. Merely as an illustration, however, of the rapidity with which the number of gardens increased during the period of which I speak, I may mention that in Italy alone gardens were founded in Turin in 1729, in Pavia and Cagliari in 1765, in Parma in 1770, in Modena in 1772, in Palermo in 1779, in Mantua in 1780, in Milan in 1781, in Siena in 1784, in Naples in 1796, in Genoa in 1803 and in Venice in 1810. And again as exemplifying the influence of the development of the artistic sense in the laying out of gardens I may ask you to contrast a photograph of the gardens of Madrid in 1781 with one which represents the same garden one hundred years later. The rectangular beds and geometrically arranged paths of the former give place to irregular figures in the ground plan of the latter; whilst even in the areas where the original ground plan has been retained the angularities have been rubbed off and a curved pathway replaces the primitive straight and formal walk. You will be able more fully to appreciate this feature in the evolution of the garden when we come to contrast modern gardens such as these of Kew, Buitenzorg and S. Louis with those of last century.

I desire now to turn for a few moments to the present day and attempt briefly to indicate to you the chief characteristics of some of the great gardens of our own times.

I have already more than once suggested to you that the botanic garden of any given period more or less reflects the condition of the science at that time. I say more or less, because science is continually progressive and botanical science more especially has, within the last fifty years advanced comparatively more rapidly than any other branch. It would be practically impossible, however, to make corresponding changes in the systematic arrangement of a garden to bring it, so to speak, up to date with the most recent research. Many no doubt extremely important investigations lead to suggestions in Taxonomy which would if followed out in practice involve enormous labour and expense. Sometimes these researches are confirmed and universally accepted and by and by they may become recognised concretely, one might say, in laboratory teaching, in museum classification and in garden arrangement. Sometimes they are not so conformed and recognised, and if the alterations in arrangement involved by their acceptance have not be carried out, there is no harm done. Imagine on the other hand the trouble and outlay rendered necessary by the renaming of the phanerograms in Kew Gardens in accordance with the views let us say of Otto Kunze, assuming that that ardent reformer's views were to receive wide acceptance, or in the transposition of even a small family of herbaceous plants from one region of the garden to another in consequence of the publication of some monograph pronouncing, even authoritatively, on its affinities. The Linnæan arrangement has of course been given up in every modern botanical garden and the natural system is to all intents and purposes universally followed. But there the similarity between gardens ends. In the first place there are in the botanic gardens in different parts of the world the necessarily different climatic and other conditions to be considered. Plants which need stove heat in Kew grow in the open air in Rio de Janeiro or Buitenzorg; plants which in our northern latitudes can be cultivated with success in open beds must be tended in frigidaria in the tropics, if they be grown at all. Moreover no two gardens, even in the same district, are quite alike. In some the principles of geographical distribution are taken as the guide in planning the garden; in others a purely systematic and taxonomic basis is selected, in others again the physiological habit, whilst others combine all three. The best gardens are those which enable the visitor to learn not only systematic or geographical or physiological botany but all of these, which are provided with museums and beds of economic and medicinal plants, with herbarium, library, and last, but by no means least, with laboratories, where in short all the manifold developments of the science during the last fifty years are represented and allowed for. There are several such gardens now in existence: time permits me to describe three only, the Missouri Garden at St. Louis, the 's Lands Plantentuin of Buitenzorg, Java, and the Royal Gardens, Kew.

With regard to the botanic gardens in the States I am indebted for much information to Mr. Coville of the U.S.

Department of Agriculture. He informs me that at the present time there are in the States but three establishments that can justly be regarded as botanical gardens in the strictest sense of the term. The first and the largest is the Missouri Garden at S. Louis, the second the Arnold Arboretum of Brookline, Mass., and thirdly the Botanic Garden at Harvard University. Another Botanic Garden Mr. Coville tells me is on the point of being established at New York City by combination of private endowment with a grant of money and land from the city itself. In the earlier history of the country there were several private gardens, such as Bartram's at Philadelphia, which is still in existence, but these cannot really be classed with the modern article. In Washington there has existed for many years a botanical garden which, however, while it contained some important and interesting collections on certain lines, has made no pretence at being a true botanic garden, and in fact has been devoted principally to ornamental purposes. The Department of Agriculture has a similar, though less complete, collection of trees and shrubs with a few herbaceous plants, and, taken in connection with the city parks, furnishes a fairly good representation of the possibilities of Washington as a centre for plant cultivation. In addition, the majority of the Colleges and Universities of the States have in connection with their botanical departments, gardens of greater or lesser extent, e.g., those of the State of Nebraska and of Cornell University, Ithaca, N.Y. The University of California has made a beginning in this direction which is likely with proper encouragement to develop into an important establishment.

In Canada also, Montreal established a botanic garden in 1886, with an area of 75 acres, chiefly at the instigation of the McGill College and the Horticultural Society of Montreal.

Henry Shaw, the founder of the great Missouri Garden and of the Shaw School of Botany, was in his early life a Sheffield cutler. He migrated to the States in 1819 and by his energy and ability ere long amassed a large fortune and retired from business whilst still in the prime of life. During a visit to the old country, in the year of the 1851 Exhibition, he spent some time in examining the more notable private gardens in England, and according to his own statement it was while walking though the grounds of Chatsworth that the idea first occurred to him to lay out a private garden in his own city of S. Louis on similar but less ambitious lines. In 1857 the Missouri Gardens were first planned on suggestions furnished him by Engelmann, Hooker, Decaisne, Alexander Braun, and other great botanists whose opinions he invited. In 1859 Shaw, at Hooker's advice, purchased the great herbarium of Professor Bernhardi of Erfurth and for the remainder of his life, indeed up to the day of his death in 1889, the garden which he had created, in the midst of which he lived and in which he now lies buried, was his constant care.

From the general plan of the garden you will readily understand the principle on which it is laid out. With the arboretum it forms a right angled triangle about 45 acres in extent, not including the adjoining meadows, which latter, though at present let, are to remain available for further extension of the gardens. The Shaw School of Botany, situated close at hand was also founded by Shaw and endowed with £1000 a year, and capital yielding a like sum was devoted to the support of a special Professor of Botany in the School, the professor being also director of the Garden. In addition to the green-houses, store-houses, and palm-house, and necessary out-buildings, a large herbarium and museum are situated within the grounds, as is also a lodge for the benefit of a number of "garden pupils" who are supported by funds left for the purpose and who are trained as scientific gardeners and foresters. The Shaw School itself has laboratories, museum and herbarium, and in the school some remarkably good work has been already accomplished, both morphological and systematic.

Elaborate in its detail and complete in its equipment as the S. Louis Garden is, it is quite eclipsed by the great tropical garden of Buitenzorg in Java, a garden which, under its distinguished director Dr. Treub, has, during the comparatively short period of its existence, done as much for the Science of Botany as any other garden in the world.

The credit of the first conception of a tropical botanic garden is due to Professor Reinwardt of Amsterdam who accompanied the Netherlands' Commissioners appointed to organise the Netherlands East Indian possessions after the termination of British rule. In 1817 Reinwardt proposed to found an experimental garden near the palace park at Buitenzorg, in Java, and his suggestion being accepted by the Dutch Government, the work was commenced that same year—a Kew gardener being appointed as curator and overseer of the works. In 1822 under Dr. Blume, the first director, the first catalogue, enumerating 900 species, was published. The succeeding twenty years was a period of anxiety for those interested in the success of the station ; economic fits on the part of the financial adminstrators and intrigues at home with the view of making the gardens a dependency of the National Museum of Leyden, and curtailing the independence and freedom of action of the garden officials, considerably retarded its development. In the struggle, fortunately for science, the colonists were triumphant, chiefly owing- to the exertions of Teijsmann, the head gardener, and the year 1866 saw the extension of the garden by the establishment of the mountain station at Tjibodas.

In 1867 Dr. Scheffer was appointed director and the following year witnessed the separation of the garden from the control of palace officialism. In 1874 the museum was built and the publication of the Annals inaugurated in the year following. In 1876 a school of Forestry and Agriculture was established, and in 1888 on the death of Scheffer the gardens entered on a period of unexampled progressive development under the directorship of Dr. Treub.

The description of the garden, as it at present stands, is rendered easy for me by the publication of the important volume "*Der Botanische Garten*" '*s Lands Plantentuin* "*zu Buitenzorg auf Java*" in 1893. In this treatise not only is there given a complete historical account of the garden but a very full description of it illustrated by many photographs several of which I shall have the pleasure of shewing you.

The garden is of irregular shape and consists of about 145 acres, of which about 30 acres are situated on an island to the eastward of the ground between two branches of the Tjilinwong, a river which bounds the garden on the east side; the north is bounded by the palace park, the west by the high road and the south by the Chinese quarter of the town of Buitenzorg. The garden is supplied with water by a tributary of the Tjilinwong which passes through the garden and feeds the lakes. The garden is intersected by numerous drives and walks. In the south west part are the residences of the director and head gardener, the offices, the laboratories, anatomical, physiological and pathological, photographic and printing rooms, the head gardener's offices, the stables, bedding and other outhouses; whilst on the south east corner are the huts of the native workmen.

The system of arrangement adopted in the garden is that of the *Genera Plantarum* of Bentham and Hooker, replacing that of Endlicher which was the system used when Teijsmann had the direction of the garden. Related families of plants are placed together in the garden, so that the system is a strictly natural one, although of course climbers, marsh plants, and such as shew peculiar habits are collected together in special regions, where their physiological characteristics may be better exhibited than by scattering them all over the garden in proximity to their various relations. The garden contains in all about 9000 species. I wished that time availed for me to give you even the faintest idea of the contents of this magnificent garden. I must leave you to imagine it from the photographs taken from the published account of the garden I have already referred to. Dr. Burck's most admirable account of the treasures of the garden described in a series of imaginary walks through the ground must be read in detail if any adequate conception is to be obtained of this the grandest botanical garden of the tropics.

The Museum originally built in 1859 to accommodate a collection of minerals illustrative of the mining districts in the Colony came into the hands of the garden officials in 1871. The present museum consists of a vestibule and large hall, round which runs a gallery and off which lead passages to various smaller rooms serving as library, keepers'-rooms, work-rooms, store-rooms and a small laboratory. The gallery is occupied by the herbarium preserved in 1200 wooden cases arranged on shelves, and includes collections made by Zollinger in the Celebes, Java and Lombock, by Teijsmann in his numerous excursions throughout the Archipelago, more especially in New

Guinea and Sumatra, and by numerous other botanists, more or less connected with the gardens, in the less known islands of Malaya. Amongst these the name of one naturalist appears, whom we are proud to honour as our vice-president—I mean Dr. H. O. Forbes. In addition to the herbarium the museum buildings contain a collection of dried fruits and seeds of tropical plants as well as a collection of flowers and fruits preserved in alcohol, an extensive series of samples of timber trees, and a large collection of vegetable products of economic importance, such as jute, hemp, india-rubber, gutta-percha, dyes, cinchona, vegetable oils, indigo, sugar, tea, coffee, rice, tobacco and the thousand and one articles of commerce of vegetable origin, known so familiarly to us in this country in their prepared state but whose native home is in the tropics.

The magnificently equipped laboratories and the wealth of tropical vegetation, literally threatening to smother the laboratories themselves, have tempted some of the most famous of our modern botanists to avail themselves of the welcome offered to all students of plant life to visit the Java Gardens by Dr. Treub and his assistants, and the researches of these savants, published in the Annals of the Buitenzorg Gardens and in various Journals and separate volumes, are amongst the most important of the many contributions to our knowledge of tropical plant life that have been published during the past decade.

Although greatly inferior to those of Buitenzorg there are many other really fine botanic gardens in the tropics. I may instance though I dare not pause to describe those of Rio de Janeiro with its stately palm avenue, of Hongkong, of Peradeniya, of Singapore and of Calcutta. In the Antipodes also Melbourne and Adelaide must be accorded a place of high rank amongst scientific Botanic Gardens.

Turning now to the gardens of Europe we may first glance at those of our own country. Unquestionably the chief of these are those of Edinburgh, Glasnevin and Kew. The Universities of Glasgow, Oxford, Cambridge and Dublin also possess gardens but of lesser size and importance. Chelsea Gardens I have already referred to.

The Edinburgh Garden was founded in 1670 to the east of the North Bridge on a site now occupied by the N.B.R. Station. In 1763 it was transferred to Leith Walk and in 1819 to Inverleith Row where it now exists and where it covers a space of about 30 acres. The Garden includes an arboretum, palm-house, hot-houses and conservatories, a herbarium, museum and class room and laboratories to accommodate the large classes of botanical students attending the University.

The Glasnevin Gardens at Dublin are much younger dating from about 1795. They were formed at the instance of the Dublin Society and were laid out and endowed by the Irish Parliament. They exceed in extent by ten acres those of Edinburgh but are very similar to them in character and plan.

No British Garden—indeed with the single exception perhaps of the great tropical garden of Buitenzorg in Java—no garden of the world approaches in extent and completeness the magnificent establishment at Kew. Its early history has been recently written by the Director, Mr. Thistleton Dyer, in the Kew Bulletin, and those who may be interested in the subject of its first beginnings I would refer to the paper in question. It will serve my present purpose if I merely state that the Gardens were established in or about 1760 by Queen Caroline, Consort of George the III., and made into a national institution in 1840 when Sir William Hooker, then professor of Botany in the University of Glasgow, was appointed

Director. The photos I am able to shew you have been specially taken for the purpose of this address and will, together with the Kew plan, enable you to gain a general notion of the Garden and its contents.

In 1840 when the garden was first made a National Institution it covered only 15 acres, now with the Arboretum it extends over 248 acres, of which 70 acres form the, garden proper. The garden is entered from Kew Green by a remarkably fine gateway, close to which are a small student's garden and the vast Herbarium, where undoubtedly there lies a more complete collection of dried plants than exists anywhere else in the world. Inside the grounds one meets first with a house devoted to tropical tree-ferns and aroids, and close at hand Museum III., the old orangery, filled chiefly with specimens of timber. A great range of houses divided into three sections occupies the centre of this section of the Gardens, and is devoted to culture of tropical and temperate ferns, succulents, heaths, begonias, victorias, temperate and tropical orchids and economic plants; there is also a store house, conservatory and alpine house. Near the rock garden and herbaceous ground is a tank for aquatics, as also Museum II. with specimens of botanical interest selected from the monocotyledonous and cryptogamic orders, whilst close to the range of bedding houses and pits, a small laboratory has been erected, conveniently situated for study and research. One cannot help wishing that the Board of Works would see its way to erecting fully equipped laboratories for anatomical and physiological research on a much larger scale. The magnificent collections at one's hand and the splendid library and herbarium in the immediate neighbourhood, encourage one to hope that at no very distinct date these advantages may be still farther added to, and the gardens made in every sense the greatest centre for botanical study and research in the world.

The Gardens are constricted about the middle by Kew Palace Grounds, and whilst the area to the north contains the buildings I have just mentioned, that to the south contains the chief museum filled with dicotyledonous examples. On the opposite side of the pond stands the palm house—a building 362 feet long, 100 feet broad and 66 feet high. The grounds surrounding all these museums and hot-houses are crowded with plants of all kinds, hardy in the open air of England. Beyond the gardens proper, is the arboretum bounded by the grounds of the Queen's Cottage, the river, the deer park and Richmond Road. The chief features of this area are the lake, the great temperate house covering three quarters of an acre and the Marianne North Gallery of paintings of indigenous vegetation in different parts of the world.

Probably there are many amongst my audience who have personally visited Kew Garden, and none who have done so can have failed to bring away with them vivid recollections of its high artistic excellence and beauty; but it needs prolonged residence and study to fully appreciate the countless botanical treasures it contains and its enormous scientific interest and value.

One word in conclusion on the gardens of the Continental Universities. Every one of them possesses a garden and although many of these are but small in area still they are rightly considered as essential adjuncts to botanical teaching. The museums are small as a rule and are obviously not considered of first-rate importance. Demonstrations are on the other hand always given in the garden and green-houses—the living plant replaces the bottled preparation. Although in our own country we of course constantly employ the garden—when we are

fortunate enough to possess one—for demonstration purposes, still I think we at the same time pay more attention
to the museum. In one respect we are immensely behind,
namely, in the extent and equipment of our botanical
laboratories: but then we in this country are not in a
position to draw freely on state aid for such trivial and
unimportant matters as University Education. What is
contributed by our Government is dealt out with par-
simonious hand and after a deal of pressing. I cannot
help thinking that if only a few of the many thousands
which are just now being annually expended by County
Councils and other public bodies on what has been termed
Technical Education were distributed amongst the Colleges
and Universities to be devoted by them, at their discretion,
to the advancement of general scientific education we
might still hope to compete successfully with our
continental colleagues. Liverpool with its three-quarters
of a million inhabitants might then be able to equip the
laboratories of its University College on a scale which
would compare favourably with, let us say, one of the
smaller German universities such as Jena or Strassburg.

I have now reached the end of my time if not of my
task.

The Botanic Garden, in the history of its development,
as I have tried, however crudely and imperfectly, to shew
you, reflects the successive phases in the advance of the
science.

What I have said of the garden is equally true of the
botanist himself in his personal training and his methods
of teaching—if he be a teacher.

In Linnaeus' time the best botanist was he who knew by
name every herb of the field—however little he might
know of each. I am not sure that the ideal aimed at
in the 18th Century has been altogether abandoned yet.

Anatomy was born, and he was considered learned who knew much about a few plants rather that little about many. The end of the 19th Century has seen the domination of yet a third school. As the systematist gave way to the anatomist so the anatomist is giving way. to the physiologist. How a plant—no matter what variety of what species it be—lives and moves and has its being—what are the functions of its various parts and how these parts work together for the common good— what part plants play in the balance of nature—in the drama of life on the earth—these are the problems that botanists attempt to solve nowadays. Doubtless it were well that every botanist should know by name every flowering—yea—every flowerless plant of the field and at the same time be conversant with its entire life cycle, its minute anatomy and the wondrous phenomena of its physiology. But *Ars longa vita brevis est*—and he has no need to be ashamed who can even say,

" *In Nature's infinite book of secrecy I can a little read.*"

APPENDIX.

The following list includes the names of the chief Botanical Gardens of the world in existence in 1895. The list does not pretend to be *absolutely* complete, although it probably includes all of scientific importance. Private gardens such as those of Tresco Abbey, La Mortola, &c., and public parks, often very extensive and not infrequently arranged on scientific lines, are not recorded. The approximate dates of foundation of some of the more important and historic gardens are inserted in brackets after the name of the city or town.

EUROPE.

AUSTRIA-HUNGARY. Agram, Budapesth, Czernowitz, Dublany, Graz, Innsbruck, Klagenfurth, Kolozsvar, Krakau, Laibach, Lemberg, Prague, Salzburg, Schemnitz, Selmeebanya, Trieste. Vienna (2).

BELGIUM. Antwerp, Anvers, Brussels, Gand, Gembloux, Liège, Louvain.

DENMARK. Copenhagen (2).

FRANCE. Angers, Besançon, Brest, Bordeaux, Caen, Cannes, Clermont-Ferrand, Dijon, Grenoble Hyères, Lille, Lyon, Marseille, Montpellier (1596), Nancy, Nantes, Orleans, Paris (4), (1597), Rochefort, Rouen, S. Quentin, Toulon, Toulouse, Tours.

GERMANY. Aachen, Bamberg, Berlin (2), (1714), Bonn, Breslau, Brunswick, Karlsruhe, Cologne, Darmstadt, Dresden, Erlangen, Strassburg (1619), Frankfurt a.M., Freiburg i.B., Giessen, Görlitz, Göttingen, Greifswald, Halle, Hamburg, Heidelberg, Jena (1629), Kiel, Königsberg, Leipzig (1580), Marburg, Münden, Munich, Munster, Potsdam, Rostock, Tharandt, Tübingen, Würzburg.

GREAT BRITAIN. London (4) [Kew (1760), Chelsea (1677), Regents Park, Horticultural Society's Garden, S.K.], Belfast, Birmingham, Cambridge, Dublin (2) [Trinity College, Glasnevin (1795)], Edinburgh (1680), Glasgow, Hull, Liverpool, Manchester, Oxford (1632).

GREECE. Athens.

HOLLAND. Amsterdam, Groningen, Leyden (1577), Utrecht, Wageningen.

ITALY. Bologna (1567), Brera, Cagliari, Camerino,

Catania, Ferrara, Florence (1550), Genoa, Lucca, Messina, Milan, Modena, Naples, Padua (1545), Palermo, Parma, Pavia, Perugia, Pisa (1547), Portici, Rome, Siena, Turin, Venice, Urbino.

MALTA.   La Valette.

NORWAY.   Christiania.

PORTUGAL.   Coimbra, Oporto, Lisbon.

ROUMANIA.   Bucharest, Yassey.

RUSSIA.   Dorpat, Helsingfors, Kasan, Kharkoff, Kiew, Moscow, Odessa, Orel, Ouman, S. Petersburg (2), Tiflis, Warsaw.

SERVIA.   Belgrade.

SPAIN.   Madrid.

SWEDEN.   Stockholm (3), Göteborg, Lund, Upsala (1660).

SWITZERLAND.   Bâle, Bern, Geneva, Lausanne, Zurich.

ASIA.

CEYLON.   Peradeniya, Hakgala, Henaratgoda, Amiràdhapura, Badulla.

CHINA.   Hongkong.

INDIA (BRITISH).   Calcutta, Mungpoo, Darjeeling, Darbhangah, Poona, Ghorpuri, Bombay, Nagpur, Ootacumund, Madras, Bangalore, Baroda, Gwalior, Morvi, Travancore, Udaipur, Agra, Allahabad, Cawnpur, Lucknow, Saharanpur Lahore, Simla.

INDIA (FRENCH).   Saïgon, Pondicherry.

JAPAN.   Tokio.

JAVA.   Buitenzorg, Tjibodas.

SIBERIA.   Tomsk.

STRAITS SETTLEMENTS.   Singapore, Penang, Malacca, Perak.

AFRICA.

Algiers, Orotava, Cairo, Cape Town, Grahamstown, Port Elizabeth, King Williamstown, Graaf Reniet, Uitenhage, Gambia, Lagos, Gold Coast, Durban, Pietermaritzburg, Old Calabar, Port Louis, S. Denis.

AMERICA.

CANADA. Montreal, Ottawa.

UNITED STATES. Brookline, Mass., Cambridge, Lansing, S. Louis, Washington, Chicago, Lincoln, Ithaca, San Francisco.

GUATEMALA. Guatemala.

WEST INDIES, &C. Barbadoes, George Town, Berbice, British Honduras, Grenada, Jamaica (6), Leewards Is. (4), S. Lucia, S. Vincent, Havana, Trinidad.

PERU. Lima.

CHILI. Santiago.

ARGENTINA. Buenos Aires.

ECUADOR. Quito.

BRAZIL. Rio de Janeiro.

AUSTRALASIA AND POLYNESIA. Sydney, Adelaide, Port Darwin, Brisbane, Rockhampton, Melbourne, Hobart Town, Wellington, Dunedin, Napier, Invercargill, Auckland, Christchurch, Honolulu.

NINTH ANNUAL REPORT of the LIVERPOOL MARINE BIOLOGY COMMITTEE and their BIOLOGICAL STATION at PORT ERIN.

By W. A. HERDMAN, D.Sc., F.R.S.,

DERBY PROFESSOR OF NATURAL HISTORY IN UNIVERSITY COLLEGE, LIVERPOOL;
CHAIRMAN OF THE LIVERPOOL MARINE BIOLOGY COMMITTEE,
AND DIRECTOR OF THE PORT ERIN STATION.

[Read 8th November, 1895.]

THE close of a third triennial period has witnessed the publication (October, 1895) of a Fourth Volume of Collected Reports by our Committee upon the Fauna of Liverpool Bay and the Irish Sea. This volume practically brings the account of the work of the Committee up to the end of the tenth year; the Committee was formed in 1885, the first volume of the "Fauna" was issued in 1886, vol. II. in 1889, vol. III. in 1892, and this fourth volume has now appeared in the autumn of 1895—giving an account of the opening of the Port Erin Station by His Excellency Dr. Spencer Walpole in 1892, and of the investigations conducted in the laboratory and at sea up to the date of our last annual report. The present (ninth) annual report brings on the record to the conclusion of the season 1895.

The Committee have carried on their usual exploring work by means of dredging expeditions and otherwise during the past year. The specimens obtained have been worked up by specialists, and some of the most noteworthy additions to our lists are given below. I am specially indebted to my colleagues on the Committee Mr. Isaac Thompson and Mr. Alfred Walker, to my Assistant Mr. Andrew Scott, and to the various other naturalists who have worked at Port Erin during the year for kind help given me in the preparation of this report.

STATION RECORD.

The following naturalists have worked at the Port Erin Laboratory during the past year :—

DATE.	NAME.	WORK.
February.	I. C. Thompson	Copepoda.
—	W. A. Herdman	Collecting.
March.	I. C. Thompson	Collecting.
—	W. A. Herdman	Collecting.
—	J. C. Sumner	Collecting.
—	R. Boyce	Collecting.
—	A. Scott	Collecting.
April.	F. G. Baily ⎫	Electric organ of Skate.
—	H. O. Forbes ⎭	
—	W. A. Herdman	Tunicata.
—	J. D. F. Gilchrist...	Opisthobranchiata.
—	J. C. Sumner	Collecting.
—	P. M. C. Kermode	Collecting.
—	A. O. Walker	Amphipoda.
May.	W. A. Herdman ⎫	Oyster experiments
—	R. Boyce ⎭	
—	P. M. C. Kermode	General.
—	J. C. Sumner	Collecting.
June.	I. C. Thompson	Copepoda
—	R. Boyce	Oysters.
—	A. Leicester	Mollusca.
—	W. A. Herdman	Tunicata.
—	A. M. Paterson	General.
—	W. I. Beaumont	Nemertines.
—	T. S. Lea	Photographing Algæ.
—	F. W. Gamble	Turbellaria.
—	J. C. Sumner	Collecting.
—	H. O. Forbes	Preserving Animals.
—	A. Scott	Copepoda, &c.
July.	W. I. Beaumont	Nemertines.
—	J. C. Sumner	Collecting.
—	T. S. Lea	Photographing Algæ.
August.	H. Meyer Delius	Studying fauna

—	J. C. Sumner	... ...	Collecting.
—	J. D. F. Gilchrist...	... ...	Mollusca.
September.	W. A. Herdman ...	... ...	General.
—	H. Meyer Delius ...	... ...	Studying fauna.
—	J. C. Sumner	... ...	Collecting.
—	F. W. Gamble	... ...	Turbellaria.
—	R. J. Harvey Gibson	... ...	Marine Algæ.
October.	I. C. Thompson ...	... ...	Copepoda.
—	R. Boyce ...	... ...	General.
—	W. A. Herdman ...	... ...	Collecting.
—	J. C. Sumner	... ...	Collecting.
November.	I. C. Thompson ...	... ...	Copepoda.
—	W. A. Herdman ...	... ...	Oyster experiments.
—	J. C. Sumner	... ...	General.

The list compares satisfactorily with those of the last few years. It shows only a slight increase in the number of workers, but some stayed for long periods, *e.g.*, Mr. Beaumont from 31st May to July 12th, Mr. Lea from June 10th to July 4th, and Mr. Delius for the two months August and September. The work done by the various naturalists at the station will be referred to further on.

THE AQUARIUM.

There is no new feature to note in connection with this part of the establishment. About 200 visitors paid for admission during the season (July and August) when it was on exhibition, while many other visitors were taken round the tanks and dishes at other times of the year when the aquarium was not formally open.

Amongst the animals which have lived in our tanks, during 1895, may be noted the angler fish (*Lophius piscatorius*), the top knot (*Zeugopterus punctatus*), the plumose anemone (*Actinoloba dianthus*) for over six months, the starfish *Solaster endeca* for over two months, the wrasse (*Labrus mixtus*), young cod and pollack, and various other fishes. Amongst other Invertebrates the

Mollusca *Doris tuberculata*, *Acanthodoris pilosa* and *Aplysia punctata* (the sea hare) spawned freely.

The basement floor of the aquarium was made use of by Professor Boyce and Professor Herdman, during a part of the summer, for some of their investigations on the life conditions and health of the oyster, and the effects of certain diseased conditions. Some further experiments on the same subject are being made in these lower tanks this winter; and the place, from its constant coolness and shade and its proximity to beach and sea, is proving admirably suited for such a purpose.

THE CURATOR.

Mr. J. C. Sumner, from the Royal College of Science, South Kensington, acted as Curator of the Biological Station from March to November, and besides his ordinary routine duties devoted much attention to improving the stock of chemicals and fixing and preserving re-agents in the laboratory. In his report to the Committee he states "I made an inventory of everything in the laboratory, all the apparatus, books, &c.; and then made a list of all the things I thought were wanted. These have been brought or sent to the station from Liverpool during the summer, so that now the place is really very well equipped......the shelves contain all the necessary fixing and killing re-agents, together with some of the commoner stains, &c." (For some faunistic notes from the Curator's diary, see p. 79). The laboratory assistant, William Bridson, is still in the employ of the Committee, at a weekly wage, and continues to give satisfaction.

TEMPERATURE OF THE SEA.

The temperatures of sea and air have not been taken with regularity through the season, but so far as the

observations go they entirely corroborate those of the
year before last which were printed in full in the Seventh.
Annual Report.   On the whole the sea off Port Erin
seems to be of a more equable temperature—slightly
warmer in winter and slightly colder in summer—than
that of the shallow waters off the Lancashire and
Cheshire coasts.

THE PROPOSED SEA-FISH HATCHERY.

It was hoped that before now some arrangement would
have been made with the Lancashire Sea-Fisheries
Committee or with the Manx Government, or with both
these bodies, whereby a Sea-Fish Hatchery for the Irish
Sea should be established at Port Erin alongside the
Biological Station.   We have now advocated that scheme
for some years, our Committee has disinterestedly offered
to assist by lending tanks for preliminary experiments, by
giving the services of their Assistant and in other ways,
and successive reports by individuals and committees
have shown that the Port Erin site is superior in natural
advantages to any of those proposed in Lancashire,
Cheshire or North Wales.   The water is pure and cool
and salt, and the configuration of shore and cliffs is such
as to lend itself readily to the formation of a large
spawning pond on the beach, while an adjacent creek
could easily be converted into a deep vivarium for lobster
culture.   Our own Committee has no funds to apply to
such a purpose, but if any of the powerful bodies interested
in promoting the fisheries of the Irish Sea, or in the
technical instruction of the fishermen, will provide the
money to erect a small experimental hatchery and spawning
pond at Port Erin, the Committee is willing to superintend
the work for the first few years, and to give time and
trouble so as to show what can be done in this locality in
the artificial cultivation of food fishes.

DREDGING EXPEDITIONS.

During 1895 the following dredging expeditions in steamers have been carried out, partly with the help, as before, of a Committee of the British Association. This B. A. Committee was re-appointed, for one year, at the Ipswich Meeting, but must bring its labours to a conclusion with a final report to the Liverpool Meeting of the Association in September 1896. With that fuller report in view for next year, the Committee do not propose now to give details* of the separate expeditions, but content themselves with the following brief summary of the occasions and localities :—

I. April 15th, 1895.—Hired steam-trawler "Lady Loch." Localities dredged, to the west and north-west of Port Erin, at depths of 20 to 40 fathoms.

II. April 25th, 1895.—Hired steam-trawler "Lady Loch." Localities dredged, to the west and south of Port Erin, at depths of 30 to 40 fathoms.

At one spot, 6 miles S.E. of Calf Island, 34 fathoms, bottom sand, gravel and shells, such a rich haul was obtained that the trawl-net tore away, and only a small part of the contents was recovered. This contained, however, a number of specimens of a rare shrimp *Pontophilus spinosus*, Leach, along with *Munida rugosa*, *Ebalia tumefacta* and *E. tuberosa*, *Xantho tuberculatus*, *Pandalus brevirostrus*, *Anapagurus hyndmanni*, *Campylaspis* sp., and *Melphidippella macera* amongst Crustaceans, and the following Echinoderma :—*Palmipes membranaceus*, *Porania pulvillus*, *Stichaster roseus*, *Luidea savignii*, *Synapta inhærens*, and other Holothurians. There were also, of course, many Mollusca, Worms, &c., and an unfamiliar Actinian, which

---

* The course of procedure on these expeditions was very fully described last year (Eighth Annual Report, p. 16) and need not be further referred to now.

Professor Haddon considers to be probably his new species *Paraphellia expansa*, previously only known from deep water off the south-west coast of Ireland.

III. June 1st, 1895.—Hired steam-trawler " Lady Loch." Localities dredged, Calf Sound and off S.E. of Isle of Man, at depths of 15 to 20 fathoms.

. IV. June 23rd, 1895.—Hired steam-trawler " Rose Ann." Localities dredged, to the W. and N.W. of Peel and Ballaugh, on the " North Bank," at depths about 20 fathoms.

V. August 3rd, 1895.—Lancashire Sea-Fisheries steamer " John Fell." Localities dredged and trawled, Red Wharf Bay and off Point Lynas, on north coast of Anglesey, at depths of 6 to 17 fathoms.

VI. August 19th, 1895.—Steamer " John Fell." Localities dredged, Carnarvon Bay, on south coast of Anglesey, depths 15 to 18 fathoms.

VII. October 27th, 1895.—Hired steamer " Rose Ann." Localities dredged and trawled, off Port Erin and along S.E. side of Island from the Calf Sound to Langness, at depths of 15 to 20 fathoms.

Additions to the Fauna.

In addition to these " steamer " expeditions there has been frequent dredging and tow-netting from small boats, and a good deal of " shore collecting."

Amongst the more noteworthy animals collected in the district during the year are the following :—

Cœlenterata.

Mr. Edward T. Browne has drawn up a list of thirty-four species of Medusæ which are found in the district, and of these the following are specially noteworthy :— *Amphicodon fritillaria* (carrying young hydroids in the umbrellar cavity), *Dysmorphosa minima*, *Cytæandra areolata* (?), *Lizzia blondina*, *Laodice calcarata* (new to European seas), and *Eutima insignis.** Mr. Browne writes, in regard to his work at the Biological Station, " Port Erin is a good place for Medusæ. The tide sweeps clean into the bay and I have found very little difference between the pelagic fauna inside the breakwater and that a mile or two off shore. At Plymouth one has to go about two miles outside the Sound before meeting the Channel tide."

Miss L. R. Thornely reports the addition of *Perigonimus repens* and *Tubiclava cornucopia* to the list of Hydroids.

VERMES.

Mr. Beaumont in his recently published report makes the following additions to the list of NEMERTIDA :— *Amphiporus pulcher*, *A. dissimulans*, *Tetrastemma flavidum*, *Prosorhochmus claparédii*, *Micrura purpurea*, *M. fasciolata*, *M. candida*, and *Cerebratulus fuscus*.

During this summer we have dredged from a gravelly bottom, at 10 to 15 fathoms, in two localities near Port Erin, a species of *Polygordius*, either P. *apogon*, M'Intosh, or a new species. It seems to differ from M'Intosh's species in having no eyes. It differs also from all the three species described by Fraipont which have no eyes.

Amongst POLYCHÆTA Mr. Sumner records *Arenicola ecaudata* and *Amphitrite johnstoni*; Mr. Arnold Watson *Autolytus alexandri* (with egg-sac), and many larval *Pectinaria*, in membranous tubes $\frac{1}{23}$ inch long.

Amongst POLYZOA Miss Thornely reports the rare *Triticella boeckii*, found attached to the prawn *Calocaris macandreæ*, from the deep mud off Port Erin ; also

---

* For Mr. Browne's observations on these and other species see his report in "Fauna of Liverpool Bay," Vol. IV., 1895.

Schizotheca divisa, Mastigophora dutertrei, Schizoporella vulgaris, and *S. cristata, Membranipora solidula, M. nodulosa,* and *M. discreta, Cribrilina gattyæ, l'orella minuta, Stomatopora incurvata,* and *Lagenipora socialis* all from the shelly deposit, at 16 to 20 fathoms, to the east of the Calf Sound.

MOLLUSCA.

The following Opisthobranchiata may be mentioned:— *Scaphander lignarius, Pleurobranchus plumula, Oscanius membranaceus, Elysia viridis, Runcina hancocki, Lamellidoris aspera, Jorunna johnstoni, Ægirus punctilucens, Polycera lessoni, Favorinus albus, Cuthona aurantiaca* and *C. nana, Coryphella gracilis, C. lineata* and *C. landsburgi, Facelina drummondi, Eolis arenicola, Cratena concinna, C. amoena* and *C. olivacea, Galvina farrani, G. tricolor* and *G. picta, Embletonia pulchra, Actæonia corrugata, Limapontia nigra, Lomanotus genei,* and a curious little *Doris,* which has been dredged several times in the neighbourhood of Port Erin, and is still unidentified. It may possibly be an unknown species. The Nudibranchs have been chiefly collected and identified by Mr. Beaumont and Mr. Sumner.

CRUSTACEA.

This section is contributed by Mr. I. C. Thompson and Mr. A. O. Walker, Mr. Thompson taking the Copepoda and Mr. Walker the higher forms. The following additional records of Copepoda have, however, been supplied by Mr. Andrew Scott independently of Mr. Thompson's report, viz.:—*Sunaristes paguri,* Hesse; *Stenhelia reflexa,* T. Scott; *Laophonte intermedia,* T. S.; *L. propinqua,* T. and A. S.; *Cletodes similis,* T. S.; *Nannopus palustris,* Brady; *Modiolicola insignis,* Aur.; and *Dermatomyzon gibberum,* T. and A. S.; all new to our fauna.

COPEPODA.

In the last report mention was made of a new copepod found by Mr. I. C. Thompson in dredged material taken outside Port Erin at 15 fathoms. This has since been described by Mr. Thompson ("Trans. Liverpool Biol. Soc.," Vol. IX., p. 26, Pls. VI. and VII.) as *Pseudocyclopia stephoides.*

It was by no means easy to decide in which genus to place this well-marked species, as it has strong points of resemblance in common with the three genera, *Pseudo-calanus, Stephos*, and *Pseudocyclopia.* With *Pseudocyclopia* it agrees in all points excepting in the number of joints in the anterior antennæ, and the primary branch of the posterior antennæ, and as in general appearance and in the first four pairs of swimming feet it strongly resembles *Pseudocyclopia* it was decided provisionally to place it in that genus. Its fifth pair of feet, however, are more like those of *Stephos.* In the "Twelfth Annual Report of the Fishery Board for Scotland" Mr. Thomas Scott added a new species belonging to this genus recently found by him in the Forth area. As the genus *Pseudocyclopia* forms a sort of missing link between the families Calanidæ and Misophriidæ, Mr. Scott has wisely constituted a new family, the Pseudocyclopiidæ, for its reception. The species of *Pseudocyclopia* described by him having respectively sixteen and seventeen joints in the anterior antennæ, he has made that number a family character. The species here described has, however, twenty joints in the anterior antennæ, and as it otherwise agrees in all respects with the family characters of Pseudocyclopiidæ Mr. Thompson suggested that the words "sixteen to seventeen jointed" be altered to "sixteen to twenty jointed" as a character of this new family, with which Mr. Scott at once concurred.

One specimen of *Modiolicola insignis*, Aurivillius, new to the district, was found in the washings of dredged material taken some miles off Peel in June, 1895. This species is known as a messmate within the shell of the "horse mussel" (*Mytilus modiolus*), and has been recorded by Canu (" Les Copep. du Boulon.," p. 238, pl. xxx., fig. 14—20), and more recently by Mr. T. Scott from the Firth of Forth. It had also been found previously by Mr. A. Scott in the "Hole" to the east of the Isle of Man.

The expedition of October 27th in the steamer "Rose Ann" was exceedingly prolific, large numbers of Copepoda being found on the bottom in shallow water (15 to 20 fathoms) although there was very little in the surface tow-nets. From some of the dredged stuff (broken shells, &c.) Mr. Scott obtained 35 species three of which, *Ameira reflexa*, *Idya gracilis* and *Tetragoniceps consimilis*, are new to the district, and eight others seem undescribed forms. Mr. Thompson has obtained already, after only a partial examination of the material, 21 species, of which *Dyspontius brevifurcatus* is new to the British fauna, and a *Cyclopicera* seems new to science. Other rich hauls still remain to be examined by Mr. Thompson.

Mr. A. O. Walker reports the following additions to our lists of the HIGHER CRUSTACEA :—

PODOPHTHALMATA.

Crangon (*Pontophilus*) *spinosus*, Leach.—Several, April 25th, 1895, station 3. Colour: whitish, freckled with reddish-brown on the antennal scales and legs; sparsely on the front and hind margins of thorax and first three abdominal segments, and densely on the last three abdominal segments, hind margin of third and generally front margin of fourth abdominal segments and proximal half of telson and lateral appendages white. Length, $2\frac{1}{4}$ in.

CUMACEA.

Hemilamprops assimilis, Sars.—Off Galley Head, Co. Cork, November 24th, 1894.

Iphinoe tenella, Sars.—Off North Bank, Peel, June 23rd, 1895. This is new to the British fauna.

Diastylis rugosoides, n.sp.—Galley Head, six males. Very near *D. rugosa* (Sars), from which it differs in the absence of the vertical plica on the carapace, and in the strong dorso-lateral teeth on the first three pleon segments.

ISOPODA.

Cirolana borealis, Lilljeborg.—Galley Head; off Port Erin, April 25th, 1895, station 2.

AMPHIPODA.

A small collection has been made by Mr. R. L. Ascroft, of Lytham, from trawl refuse and a tow-net attached to the trawl beam when working in the southern part of the Irish Sea off Galley Head. The most interesting feature of it is that nearly all the specimens are adult males, in which condition amphipods are less often taken than any other. This may perhaps be attributed to their having been taken late in November, a season at which collectors do not generally dredge.

Parathemisto oblivia, Kröyer.—Galley Head.

Callisoma crenata, Bate.—Galley Head; off Port Erin, April 25th, 1895, station 1.

Hippomedon denticulatus, Bate.—Galley Head.

Orchomenella ciliata, Sars.—Galley Head.

Tryphosites longipes, Bate.—Galley Head.

Lepidepecreum carinatum, Bate.—Galley Head.

Paraphoxus oculatus, Sars.—Off Port Erin, April 25th, 1895, stations 1 and 2.

* Those species marked with a star are new to the British fauna.

Epimeria cornigera, Febr.—Galley Head.

Syrrhoë fimbriata, Stebbing and Robertson.—Off Port April 25th, 1895, station 1.

Leptocheirus hirsutimanus (Bate) = *L. pilosus,* Sars, not Zaddach.—Two miles south-east of Kitterland, 17 fathoms, May 27th, 1894.

Photis longicaudatus, Bate.—Off Port Erin, April 25th, 1895, stations 2 and 3.

**Photis pollex,* n.sp.—Colwyn Bay, shore; Little Orme; Menai Straits, 5 to 10 fathoms. This species is intermediate between *Photis reinhardi* (Kröyer) and *P. tenuicornis* (Sars). The hind margin of the propodos of the second gnathopod in the male is distally produced into a thumb-like process which has its origin much nearer the carpus than in *P. reinhardi.*

Podocerus ocius, Bate.—Sponge *débris,* Port Erin, 1894.

PYCNOGONIDA.

The following rare species found during the year at Port Erin have been named by Mr. G. H. Carpenter, of Dublin:—*Anoplodactylus petiolatus,* Kr., *Ammothea echinata,* Hodge, *Nymphon gracile,* Leach, *N. gallicum,* Hoek, *Chætonymphon hirtum,* Kr., and *Pallene producta,* Sars, the last apparently new to Britain.

———

SOME STATISTICS OF DREDGING RESULTS.

During this year's work we have been paying some attention to the actual numbers of individuals, species, and genera brought up in particular hauls of the dredge or trawl. Our attention has recently been directed to the matter by some statements in Dr. Murray's summary volumes of the "Challenger" Expedition Report which seemed not to be quite in accord with our own experience. Dr. Murray quotes the statistics of the Scottish Sea-

Fisheries Board to show that only 7·3 species of invertebrates and 8·3 species of fishes are captured on the
average by the Fisheries steamer "Garland's" beam
trawl; and he cites as an example of a large and varied
haul from deep water one taken by the "Challenger" at
station 146 in the Southern Ocean, at a depth of 1,375
fathoms, with a 10-foot trawl dragged for at most 2 miles
during at most two hours, when 200 specimens were
captured belonging to fifty-nine genera and seventy-eight
species.   Murray then goes on to say: "In depths less
than 50 fathoms, on the other hand, I cannot find in all
my experiments any record of such a variety of organisms
in any single haul, even when using much larger trawls
and dragging over much greater distances."   Now our
experience of dredging in the Irish Sea is that quite
ordinary hauls of the dredge or very small trawl (only 4-foot
beam) contain often more specimens, species, and genera
than the special case cited from the "Challenger" results.
On the first of our expeditions after the appearance of Dr.
Murray's volumes we counted the contents of the first
haul of the trawl.   The particulars are as follows :—June
23rd, 7 miles W. of Peel, on North Bank, bottom sand
and shells, depth 21 fathoms, trawl 4 feet beam, down for
20 minutes; 232 specimens were counted, but there may
well have been another 100; they belonged to at least 112*
species and 103 genera, a larger number in every respect—
specimens, species, and genera—than in the "Challenger"
haul quoted.   The list of these species is here given, and
marine zoologists will see at a glance that it is nothing
out of the way, but a fairly ordinary assemblage of not
uncommon animals such as is frequently met when
dredging in from 15 to 30 fathoms.

---

* Really an under estimate, several other species have been identified
since from the same haul.

SPONGES:
  Reniera, sp.
  Halichondria, sp.
  Cliona celata
  Suberites domuncula
  Chalina oculata
CŒLENTERATA:
  Dicoryne conferta
  Halecium halecinum
  Sertularia abietina
  Coppinia arcta
  Hydrallmania falcata
  Campanularia verticillata
  Lafoëa dumosa
  Antennularia ramosa
  Alcyonium digitatum
  Virgularia mirabilis
  Sarcodictyon catenata
  Sagartia, sp.
  Adamsia palliata
ECHINODERMATA:
  Cucumaria, sp.
  Thyone fusus
  Asterias rubens
  Solaster papposus
  Stichaster roseus
  Porania pulvillus
  Palmipes placenta
  Ophiocoma nigra
  Ophiothrix fragilis
  Amphiura chiajii ˙
  Ophioglypha ciliata
  O. albida
  Echinus sphæra

  Spatangus purpureus
  Echinocardium cordatum
  Brissopsis lyrifera
  Echinocyamus pusillus
VERMES:
  Nemertes neesii
  Chætopterus, sp.
  Spirorbis, sp..
  Serpula, sp.
  Sabella, sp.
  Owenia filiformis
  Aphrodite aculeata
  Polynoe, sp.
CRUSTACEA:
  Scalpellum vulgare
  Balanus, sp.
  Cyclopicera nigripes
  Acontiophorus elongatus
  Artotrogus magniceps
  Dyspontius striatus
  Zaus goodsiri
  Laophonte thoracica
  Stenhelia reflexa
  Lichomolgus forficula
  Anonyx, sp.
  Galathea intermedia
  Munida bamffica
  Crangon spinosus
  Stenorhynchus rostratus
  Inachus dorsettensis
  Hyas coarctatus
  Xantho tuberculatus
  Portunus pusillus
  Eupagurus bernhardus

E. prideauxii
E. cuanensis
Eurynome aspera
Ebalia tuberosa
POLYZOA:
Pedicellina cernua
Tubulipora, sp.
Crisia cornuta
Cellepora pumicosa and three
 or four undetermined spec-
 ies of Lepralids
Flustra securifrons
Scrupocellaria reptans
Cellularia fistulosa
MOLLUSCA:
Anomia ephippium
Ostrea edulis
Pecten maximus
P. *opercularis*
P. *tigrinus*
P. *pusio*
Mytilus modiolus
Nucula nucleus
Cardium echinatum
Lissocardium norvegicum

Solen pellucidus
Venus gallina
Lyonsia norvegica
Scrobicularia prismatica
Astarte sulcata
Modiolaria marmorata
Saxicava rugosa
Cyprina islandica
Chiton, sp.
Dentalium entale
Emarginula fissura
Velutina lævigata
Turritella terebra
Natica alderi
Fusus antiquus
Aporrhais pes-pelicani
Oscanius membranaceus
Doris, sp.
Coryphella landsburgi
Tritonia plebeia
TUNICATA:
Ascidiella virginea
Styelopsis grossularia
Eugyra glutinans
Botryllus, sp.
B., sp.

The following are two other similar hauls taken with different instruments (dredge and trawl), but both in less than 20 fathoms. On October 27th, 1895, in the steam-trawler "Rose Ann" we counted the first haul of the dredge (2 feet of scraping edge) and the first haul of the small trawl (4 foot beam) with the following results:—

First haul of dredge, across mouth of Port Erin Bay, from Bradda Head towards the Calf Sound, depth 17 fathoms, bottom dead shells, 93 species in 81 genera.

Ascetta primordialis
Cliona celata
Halecium halecinum
Sertularella polyzonias
Hydrallmania falcata
Antennularia antennina
Lafoea dumosa
Obelia, sp.
Asterias rubens
Henricia sanguinolenta
Solaster papposus
Ophiothrix fragilis
Echinus sphæra
Polynoe, sp.
Serpula, sp.
Pomatoceros triqueter
Spirorbis, sp.
Terebella nebulosa
Mucronella peachii
M. ventricosa
Smittia reticulata
Membranipora craticula
M. flemingii
M. dumerilii
M. imbellis
Microporella malusii
M. ciliata
Lichenopora hispida
Schizoporella linearis
S. hyalina
Idmonea serpens
Scrupocellaria reptans
Tubulipora flabellaris
Crisia, sp.

Diastopora suborbicularis
D. patina
Porella concinna
Chorizopora brongniartii
Cellepora costazii
Balanus balanoides
Chthamalus stellatus
Cyclopina gracilis
Misophria pallida
Thalestris clausii
Ectinosoma spinipes
Cyclopicera lata
C. nigripes
Lichomolgus maximus
Dermatomyzon gibberum
Artotrogus magniceps
Zaus goodsiri
Iphimedia obesa
Melita obtusata
Lilljeborgia kinahani
Aora gracilis
Erichthonius abditis
Phtisica marina
Gnathia (Anceus), sp.
Hyas araneus.
H. coarctatus
Hippolyte varians
Spirontocaris spinus
Eupagurus bernhardus
Galathea intermedia
Ebalia tuberosa
Portunus, sp.
Achelia echinata
Anomia ephippium

Nucula nucleus

Mytilus modiolus

Pecten opercularis

P. maximus

P. pusio

Saxicava rugosa

Venus lincta

Tapes, sp.

Cyprina islandica

Chiton, sp.

Emarginula fissura

Velutina lævigata

Capulus hungaricus

Buccinum undatum

Fusus antiquus

Trochus cinerarius

Eolis viridis

Polycera quadrilineata

Perophora listeri

Ciona intestinalis

Ascidiella virginea

Ascidia mentula

A. scabra

Styelopsis grossularia

Cynthia morus

The first haul of the small trawl, on the same occasion, off the Halfway Rock, in 18 fathoms yielded 111 species in 93 genera, as follows :—

Leucosolenia coriacea

Suberites domuncula

Cliona celata

Coppinia arcta

Sertularia abietina

Antennularia ramosa

Plumularia, sp.

Sagartia nivea

Sarcodictyon catenata

Palmipes membranaceus

Solaster endeca

Asterias rubens

Henricia sanguinolenta

Porania pulvillus

Echinus sphæra

Echinocyamus pusillus

Lineus marinus

Amphiporus pulcher

Micrura fasciolata

Filograna implexa

Serpula, sp.

Pomatoceros triqueter

Polynoe, sp.

Ætea recta

Scrupocellaria scrupea

S. reptans

Idmonea serpens

Schizotheca fissa

Crisia ramulosa

C. cornuta

Cellepora pumicosa

C. dichotoma

Alcyonidium gelatinosum

A. mytili

Cellaria fistulosa

Membranipora pilosa

M. craticula

M. flemingii

M. imbellis

Chorizopora brongniartii

Smittia trispinosa

S. reticulata

Schizoporella linearis

Mucronella peachii

M. ventricosa

M. coccinea

Porella concinna

Diastopora obelia

Microporella malusii

Hippothoa divaricata

H. distans

Stomatopora johnstoni

Balanus balanoides

Thalestris peltata

Dactylopus flavus

Laophonte spinosa (?)

Ectinosoma atlanticum

Cyclopicera gracilicauda

Lichomolgus liber

Dyspontius striatus

Acontiophorus scutatus

Artotrogus orbicularis

Stenothoe marina

Leucothoe spinicarpa

Amphilochus manudens

Cyproidea brevirostris

Tritæta gibbosa

Cressa dubia

Podocerus cumbrensis

Spirontocaris spinus

Stenorhynchus, sp.

Portunus, sp.

Eupagurus bernhardus

E. cuanensis

Galathea intermedia

G. dispersa

Pandalus annulicornis

Crangon allmani

Xantho tuberculatus

Pycnogonum littorale

Anomia ephippium

Ostrea edulis

Mytilus modiolus

Pecten maximus

P. tigrinus

P. pusio

P. opercularis

Astarte, sp.

Venus casina

Tapes, sp.

Nucula nucleus

Saxicava rugosa

Pectunculus glycimeris

Chiton, sp.

Cyprina islandica

Tectura virginea

Emarginula fissura

Pleurotoma, sp.

Trochus millegranus

T. zizyphinus

Goniodoris nodosa

Amaroucium, sp.

Didemnum, sp.

Leptoclinum maculatum

Botryllus, sp.

Ascidiella virginea

Ascidia mentula

A. plebeia

Corella parallelogramma

Styelopsis grossularia

Cynthia morus

A third haul, on this same occasion (October 27th) gave us, from 16 fathoms, 156 species (see below, p. 64).

In order to get another case, on entirely different ground, not of our own choosing, on the first occasion after the publication of Dr. Murray's volumes when we were out witnessing the trawling observations of the Lancashire Sea-Fisheries steamer "John Fell," I counted, with the help of Mr. Andrew Scott and the men on board, the results of the first haul of the shrimp trawl. It was taken on July 23rd at the mouth of the Mersey estuary, inside the Liverpool Bar, on very unfavourable ground : bottom muddy sand, depth 6 fathoms. The shrimp trawl (1½-inch mesh) was down for 1 hour, and it brought up over seventeen thousand specimens referable to the following thirty-nine species belonging to thirty-four genera :—

Solea vulgaris

Pleuronectes platessa

P.　limanda

Gadus morrhua

G.　æglefinus

G.　merlangus

Clupea spratta

C.　harengus

Trachinus vipera

Agonus cataphractus

Gobius minutus

Raia clavata

R.　maculata

Mytilus edulis

Tellina tenuis

Mactra stultorum

Fusus antiquus

Carcinus mænas

Portunus, sp.

Pagurus bernhardus

Crangon vulgaris

Sacculina, sp.

Amphipoda (undetermined)

Longipedia coronata

Ectinosoma spinipes

Sunaristes paguri

Dactylopus rostratus

Cletodes limicola

Caligus, sp.

Flustra foliacea

Aphrodite aculeata

Pectinaria belgica

Nereis, sp.

Asterias rubens

Hydractinia echinata

Sertularia abietina

Hydrallmania falcata

Aurelia aurita

Cyanæa, sp.

These numbers have been exceeded on many other hauls in the ordinary course of work by the Fisheries steamer in Liverpool Bay. For example, on this occasion the fish numbered 5,943, and I have records of hauls in which the fish numbered over 20,000. The shrimps probably number as many again, and if the starfishes and other abundant invertebrates are added the total must sometimes reach such enormous numbers as from 45,000 to 50,000 specimens in a single haul of the trawl in shallow water, not including microscopic forms. Hauls such as this are doubtless as prolific of *individuals* as any of those hauls sometimes quoted containing large numbers of specimens (*of a very few species*) of Copepoda and Schizopoda from waters deeper than 50 fathoms,* and are certainly far more prolific in species and genera ; while hauls such as the three quoted above under dates June 23rd and October 27th compare favourably as to variety of life, *i.e.*, as to number of *species* and *genera*, with the deep water hauls of the " Challenger " expedition made with a far larger trawl.

On the next occasion, when on board the " John Fell," on our own expedition of August 3rd, two members of this Committee (A. O. Walker and W. A. Herdman) identified the species brought up in the first haul of the

---

* Such as those referred to by Mr. Turbyne in " Nature " for October 24th, 1895, which illustrate an interesting case of distribution of a very few species but do not affect the argument given above for the relative richness, haul for haul, of the shallow as compared with the deeper waters.

trawl (5-inch mesh), taken in Red Wharf Bay, Anglesey, at a depth of 4 to 7 fathoms. They were 78 species, belonging to 67 genera, as follows :—

Solea vulgaris
S. lutea
Pleuronectes platessa
P. limanda
P. flesus
Gadus morrhua
G. æglefinus
G. merlangus
Callionymus lyra
Raia maculata
Fusus antiquus
Buccinum undatum
Natica alderi
Pleurotoma, sp.
Philine, sp.
Eolis, sp.
Polycera quadrilineata
Corbula gibba
Mactra stultorum
Scrobicularia alba
Portunus depurator
Corystes cassivelaunus
Hyas coarctatus
Stenorhynchus phalangium
Eupagurus bernhardus
Crangon vulgaris
Pseudocuma cercaria
Diastylis rathkei
D. spinosa
Balanus balanoides
Paratylus swammerdammii

Harpinia neglecta
Ampelisca lævigata
Monoculodes longimanus
Amphilochus melanops
Pariambus typicus
Achelia echinata
Aphrodite aculeata
Nereis, sp.
Terebella, sp.
(?) Syllis, sp.
Serpula, sp.
Spirorbis, sp.
Cellaria fistulosa
Flustra foliacea
Eucratea chelata
Scrupocellaria reptans
Bugula, sp.
Cellepora pumicosa
C. avicularis
Porella compressa
Mucronella peachii
Membranipora membranacea
M. pilosa
Alcyonidium gelatinosum
Vesicularia spinosa
Gemellaria loricata
Lichenopora hispida
Crisia eburnea
C. cornuta
Idmonea serpens
Asterias rubens

Amphiura squamata	*Antennularia ramosa*
Ophioglypha albida	*Coppinia arcta*
Tealia crassicornis	*Sertularella polyzonias*
Alcyonium digitatum	*Sertularia abietina*
Clytia johnstoni	*S. argentea*
Lafoëa dumosa	*Diphasia rosacea*
Hydrallmania falcata	*D. tamarisca*
Halecium halecinum	*Tubularia indivisa*

This was a haul—from very shallow water—which combined mere quantity of life, *i.e.*, number of *individuals*, with variety of life or *number of species and genera.* The ten species of fish were represented by 879 individuals, and we estimated that there were some hundreds of crabs and of starfishes, and some thousands of shrimps. The numbers of the Molluscs, of the hermit-crabs, of *Balanus* and of *Spirorbis* were also very large.

From these statements it is clear that whether it be a question of mere *mass* of life or of *variety* of life, haul for haul, the shallow waters can hold their own against the deep sea, and form in all probability the most prolific zone of life on this globe.

Relations of Genera to Species.

A point which comes out in making complete lists, such as those given above, of the contents of the net on one haul is the relatively large number of genera represented by the species.* In the haul, quoted above, from the expedition of June 23rd, the 112 species were referred to 103 genera; in the haul from the Fisheries steamer on July 23rd, the 39 species obtained belong to 34 genera; on August 3rd, there were 78 species and 67 genera, and

---

* Dr. Murray, in the Challenger "Summary," notes this fact in the case of deep-sea hauls, but does not seem to recognise its application to shallower waters.

in the two hauls of October 27th there were 93 species in 81 genera and 111 species in 93 genera. Taking a few instances of particular groups—on August 25th, 1894, the 15 species of Tunicata taken in one haul represented 10 genera; and Mr. Walker reports the following numbers of species and genera in hauls of the higher Crustacea :— March, 1893, off Rhos, shallow, 19 species in 18 genera ; May, 1893, off Rhos, two fathoms, 24 species in 21 genera ; July, 1893, off Little Orme, 5 to 10 fathoms, 31 species in 28 genera; October, 1893, off Little Orme, 4 to 10 fathoms, 41 species in 36 genera; September, 1894, off Little Orme, shallow, 39 species in 35 genera ; and April, 1895, off Port Erin, 34 fathoms, 40 species in 35 genera.*

These figures are particularly interesting in their bearing on the Darwinian principle that an animal's most potent enemies are its own close allies.† Is it then the case, as the above cited instances suggest, that the species of a genus rarely live together; that if in a haul you get half-a-dozen species of lamellibranchs, amphipods, or annelids they will probably belong to as many genera, and if these genera contain other British species these will probably occur in some other locality, perhaps on a different bottom, or at a greater depth? It is obviously necessary to count the total number of genera and species of· the groups in the local fauna, as known, and compare these with the numbers obtained in particular hauls. That has been

* These numbers refer to the Higher Crustacea only. There were many other animals in the hauls.

† "As the species of the same genus usually have, though by no means invariably, much similarity in habits and constitution, and always in structure, the struggle will generally be more severe between them, if they come into competition with each other, than between the species of distinct genera." Darwin, The Origin of Species, sixth edition, p. 59 ; see also Wallace, Darwinism, second edition, p. 33.

done to some extent with the "Fauna" of Liverpool Bay, and the following instances may be taken as samples.

The known number of species of higher Crustacea in Vol. I. of the "Fauna" (1886) is 90, and these fall into 60 genera. But many species have been added since then, so Mr. Walker has gone over the records up to date (1895), and states that we now know in our local fauna 230 species which belong to 150 genera. This is still much the same proportion as in the former numbers, so we may take it that in our district the genera are to the species as 2 to 3, whereas in the collections quoted from Mr. Walker above the genera are to the species on the average about as 28 to 31, or nearly 7 to 8. It can also be brought out by similar series of numbers that as one extends the area investigated the number of species per genus is increased. In a single haul, in our district as we have seen, the species are to the genera about as 8 to 7, in our local fauna the proportion is about 3 species to 2 genera, while in the much wider area embraced by Sars' Amphipoda of Norway the numbers are 365 species to 157 genera or nearly 5 species to 2 genera. In other words if allied species, taking a large district, were associated together we might expect to find about twice as many species per genus in each haul as we do find.

Mr. Walker has gone carefully into this matter of the proportion of genera to species in our hauls, and in other areas, and from the figures in his notes I extract the following records, in support of the above statement :—

Rhos Bay, 13/5/93, Amphipoda, 16 sp. in 14 genera.

Little Orme, 28/7/93, Amphipoda, 24 sp. in 22 genera.

      ,,   5/10/93,   ,,   29 ,, 24 ,,

The aggregate of the species and genera of Amphipoda in the above three dredgings is 69 species in 60 genera, or an average proportion of 115 : 100. Now the total number of Amphipoda so far recorded in the L. M. B. C. District (about 5000 sq. miles) is 124 species in 78 genera, or in the proportion of 159 : 100 ; while G. O. Sars dealing with the Amphipoda of Norway—a very much more extended area—gives 365 species in 157 genera, or in the proportion of 232 : 100.

To sum up, the proportions of species and genera, in these Amphipoda are :—

Rhos, &c., 1 to 10 fms., 115 sp. in 100 gen., or $1\frac{1}{6}$: 1.

L.M.B.C. dist., to 70 fms., 159 sp. in 100 gen., over $1\frac{1}{2}$: 1.

Norway, to 1215 fms., 232 sp. in 100 gen., or $2\frac{1}{3}$: 1.

So, it is clear that as one increases the area and depth investigated the proportion of species to genera in the fauna increases, until *e.g.*, on the coasts of Norway it has become more than twice what it is on the north coast of Wales.

Again, the total number of recorded species of L.M.B.C. Tunicata is 46, and these are referred to 20 genera ; while in the case given above (August 25th, 1894) the 12 species taken on one spot represented 10 genera, or, a little over a quarter of the species represented half the genera. These and many other series of statistics in regard to other groups which we might quote, seem to show that a disproportionately large number of genera is represented by the assemblage of species at one spot, which means that closely related species are, as a rule, not found together.

We know of some individual cases, of course, of allied species occurring together, but these do not necessarily affect the general argument. Exceptional cases may be due to some special habit which, although the species are allied forms, prevents them from being severe competitors. It is possible also that sessile animals, such as hydroids and polyzoa, may form a partial exception, and may differ from wandering forms in˙ their method of competition. However Miss Thornely finds that in most gatherings of Polyzoa the species are less than twice the number of genera, while in our "Fauna" the average recorded number is 2·5 species in a genus. Moreover the colonies on dead shells or on stones are generally not only distinct species, but also distinct genera. As many as ten genera are sometimes represented by the Polyzoon colonies on one shell. We are accumulating further statistics on all these points.

The Submarine Deposits.

In last year's report the nature of the deposits forming on the floor of the Irish Sea was discussed in a preliminary manner. During this season's work the bottom brought up on each occasion has been carefully noted and a sample kept for future study in the Jermyn Street Museum. One point which this collection of deposits from comparatively shallow shore waters seems to bring out is that the classification of submarine deposits into "terrigenous" and "pelagic," which was one of the earliest oceanographic results of the " Challenger " Expedition, and which is still adhered to in the latest " Challenger " volumes as an accepted classification, does not adequately represent or express fully the facts. Terrigenous deposits are supposed to be those formed round continents from the waste of the land, and are stated to contain on the average 68 per cent. of silica. Pelagic deposits are those formed in the open ocean from the shells and other remains of animals and plants living on the surface of the sea above, and they are almost wholly free from quartz particles.

Ordinary coast sands and gravels and muds are undoubted *terrigenous* deposits. Globigerina and Radiolarian oozes are typical *pelagic* deposits. But in our dredgings in the Irish Sea, where the deposits ought all, from their position, to be purely terrigenous, we meet with several distinct varieties of sea-bottom which are not formed mostly from the waste of the land, and do not contain anything like 68 per cent. of silica; but, on the contrary, are formed very largely of the remains of bottom haunting plants and animals, and may contain as little as 17 per cent. of silica. Such are the nullipore bottoms, and the shell sand and shell gravel met with in some places, and the sand formed of comminuted spines and plates of echinoids which we have found off the Calf Island. These deposits are really much more nearly allied in their nature, and in respect of the kind of rock which they would probably form if consolidated,* to the calcareous oozes amongst pelagic deposits, than they are to terrigenous deposits, and yet they are formed on a continental area close to land in shallow water. Moreover, although agreeing with the pelagic deposits in being largely organic in origin, they differ in being derived not from surface organisms, but from plants (the nullipores) and animals which lived on the bottom. Consequently the division of deposits into "terrigenous" and "pelagic" ought to be modified or replaced by the following classification:—

1. *Terrigenous* (Murray's term, restricted)—where the deposit is formed chiefly (say, at least two-thirds, 66 %) of mineral particles derived from the waste of the land.

2. *Neritic†*—where the deposit is largely of organic origin, its calcareous matter being

---

* They seem closely comparable with the Coralline and Red Crag formations of Suffolk.

† Adopted from Haeckel's term for the zone of shallow water marine fauna, see "*Plankton Studien*," Jena, 1890; also Hickson's "Fauna of Deep Sea," 1894.

                derived from the shells and other hard parts of the animals and plants living on the bottom.

3. *Planktonic* (Murray's pelagic)—where the greater part of the deposit is formed of the remains of free-swimming animals and plants which lived in the sea above the deposit.

The last group is Murray's " pelagic " unchanged, and *that,* there can be no doubt, is a natural group of deposits; but Murray's " terrigenous " is an unnatural or hetero-geneous assemblage containing some deposits, such as the gravel off Bradda Head and the sand of the Liverpool bar, which are clearly terrigenous in their origin, along with others such as shelly sands and nullipore deposits which have much less to do with the waste of the land, but are very largely organic in origin and formed by animals and plants *in situ.* The proposal is then to recognise this latter group of deposits by separating them from the truly terrigenous under the name " Neritic." Probably some of the Coral sands described by Murray and Renard in their Challenger Report on Sea-Deposits would also fall into this category.

Professor Johannes Walther, of Jena, who has of recent years been working on the borderlands of geology and bionomics, in a recent letter to me on my proposed classification of deposits says :—" Ich meine dass der Ausdruck *benthonisch* statt *neritisch* richtiger wäre. Denn es kommt doch bei der Diagnose weniger daraufan, dass die Ablagerung in der Flachsee, als dass sie durch *benthonisch Organismen* (Coralline, Korallen, Echinoder-men, Mollusken, Bryozoen, etc.) gebildet wird." With this I can quite agree. I lay most stress on the nature (bottom plants and animals) of the particles composing the deposits, and I do not mind much whether they are called Neritic or Benthonic so long as the category is recognised as distinct from terrigenous.

Dr. C. Kohn has kindly analysed for me a series of fair samples of deposits from different parts of the Irish Sea, with the following results:—

| | NERITIC. | | | | TERRIGENOUS. | | | |
	A.*	B.†	C.	D.	E.	F.	G.	H.
Silica. SiO².	16·83	46·65	54·84	23·41	84·62	83·06	77·10	78·92
Cal. Carbonate. CaCO³.	79·27	38·45	39 71	59·66	6·38	9·18	9·20	8·59
Residue (other than Silica).	3·90	14·90	6·45	19·93	9·00	13·70	13·70	12·69
	100·00	100·00	100·00	100·00	100·00	100·00	100·00	100·00

The localities and particulars are,—

A. 1 mile off Spanish Head, 16 faths., shell fragments.

B. 1 mile off Calf of Man, 20 faths., shells and spines.

C. 1 mile off Calf of Man, 18 faths., shell sand, spines.

D. 2 miles off Dalby, 15 faths., nullipores.

E. Liverpool Bar, 3 faths., sand.

F. Bahama Bank, 13 faths., muddy sand.

G. King William Bank, 5 faths., coarse sand.

H. North end of "Hole," 28 faths., mud.

It will be noticed that the four terrigenous deposits (sands and muds) all show less than 10 % of calcium carbonate; while the four neritic have all more than 38 %—well over a third—of calcium carbonate, and one (A.) has over 79 %. The silica in these neritic deposits may be less than 17 %,

---

* Shelly deposit. Contained 1·09 % of small stones not included in analysis.

† Contained 4·82 % of magnesium carbonate, in addition to calcium carbonate.

and does not rise in any to 55 %. In round numbers it may be said that in these examples the silica makes up from 20 to 50 % and the calcium carbonate from 40 to 80 %. In all the neritic deposits there are in the residue small quantities of calcium phosphate, of iron and of alumina.

In some of these deposits the calcareous matter is formed almost entirely of *Lithothamnion*. Amongst the Nullipores from our neritic deposits Professor Harvey Gibson has identified the following species:—*Lithothamnion polymorphum*, *L. calcareum*, *L. agariciforme*, *L. fasciculatum*, the variety *fruticulosum*, and *Lithophyllum lenormandi*.

One of these neritic deposits (A) has its calcareous matter formed by a large number of animals, belonging to various groups, in addition to nullipores. One sample (measuring 7 qts., 1½ pts., and weighing, when dry, 17 lbs. 3¼ ozs.) which I have gone over carefully for the purpose of identifying the constituent particles contains more or less fragmentary remains of at least the following 99 species, all of them forms that leave calcareous remains :—

NULLIPORES :
Lithothamnion fasciculatum
L. calcareum
Lithophyllum lenormandi
ECHINODERMATA :
Echinus sphæra
Echinocyamus pusillus
Echinocardium cordatum
VERMES :
Serpula, sp.
Spirorbis, sp.
POLYZOA :†
Cellaria fistulosa

Cellepora avicularis
C. dichotoma
C. pumicosa
Idmonea serpens
Microporella ciliata
M. malusii
M. violacea
Schizotheca fissa
*S. divisa
Mastigophora hyndmanni
*M. dutertrei
Mucronella peachii
M. variolosa

* New to district.   † Identified by Miss L. R. Thornely.

M. ventricosa
M. coccinea
*Do., var. mamillata
Schizoporella auriculata
S. linearis
S. unicornis
S. simplex
*S. vulgaris
S. discoidea
*S. cristata
Membranipora catenularia
M. solidula
M. pilosa
*M. nodulosa
*M. discreta
M. aurita
M. craticula
M. flemingii
Hippothoa distans
Cribrilina annulata
C. punctata
*C. gattyæ
Crisia aculeata
Ætea recta
Amathia lendigera
Porella concinna
Do., var. belli
*P. minuta
Cribrilina radiata
Phylactella collaris
P. labrosum
Micropora coriacea
Chorizopora brongniartii
Diastopora patina

D. obelia
D. suborbicularis
Stomatopora johnstoni
*S. incurvata
Lepralia foliacea
L. pertusa
Lagenipora socialis
Smittia trispinosa
S. reticulata
Lichenopora hispida
CRUSTACEA :
Balanus balanoides
Verruca, sp.
Cancer pagurus
MOLLUSCA :
Anomia ephippium
Lima elliptica
Pecten opercularis
P. pusio
Cardium edule
Venus casina
V. ovata
Nucula nucleus
Mytilus edulis
Saxicava rugosa
Tapes, sp.
Mactra solida
Pectunculus glycimeris
Acmæa testudinalis
A. virginea
Emarginula fissura
Helcion pellucidum
Trochus magus
T. millegranus

T. cinerarius
Pleurotoma, sp.
P., sp.
Murex erinaceus
Phasianella pullus
Natica, sp.

Buccinum undatum
Capulus hungaricus
Cypræa europea
Nassa incrassata
Rissoa, sp.
R., sp.

From a bag of this shelly Neritic deposit (A), described above, Mr. Andrew Scott has by careful examination managed to extract the following 36 species of Copepoda, of which 4 are new records for our district and 8 others seem new to science:—*Pseudocyclops obtusatus*, B. & R.; *Ectinosoma sarsii*, Boeck; *E. melaniceps*, Boeck; *E. erythrops*, Brady; *E. gracile*, T. & A. Scott; *Tachidius brevicornis*, Müller; *Stenhelia*, n.sp.; *Stenhelia*, n.sp.; *Ameira longipes*, Brady; *A. longicaudata*, T. Scott; *A. reflexa*, T. Scott; *A. gracile*, n.sp.; *Mesochra macintoshi*, T. & A. S.; *Paramesochra dubia*, T. Scott; *Tetragoniceps consimilis*, T. Scott; *Laophonte thoracica*, Boeck; *L. curticaudata*, Boeck; *Pseudolaophonte aculeata*, n.gen. and n.sp.; *Normanella attenuata*, n.sp.; *Dactylopus stromii*, Baird; *D. tenuiremis*, B. & R.; *D. flavus*, Claus; *D. brevicornis*, Claus; *Thalestris rufocincta*, Norman; *T. peltata*, Boeck; *Harpacticus chelifer*, Müller; *Zaus spinatus*, Goodsir; *Z. goodsiri*, Brady; *Idya gracilis*, T. Scott; *Lichomolgus fucicolus*, Brady; *L. furcillatus*, Thorell; *Dermatomyzon nigripes* (B. & R.); *Ascomyzon thompsoni*, n.sp.; *Acontiophorus scutatus*, B. & R.; and two other species which have not yet been worked out.

Mr. Thompson has also identified from a sample of the same deposit which he examined a number of the above species, and in addition the following five:—*Porcellidium*, sp., *Ameira attenuata*, *Laophonte spinosa*, *Scutellidium fasciatum*, and *Artotrogus orbicularis*, making 41 species of Copepoda in all.

These 41 species added to the 99 species from the same haul noted on p. 64 and to the following 16 species recorded from the trawl on Oct. 27th, when the haul was taken, make in all 156 species :—*Mytilus modiolus, Pecten tigrinus, Trochus zizyphinus, Fissurella græca, Eulima polita, Pagurus prideauxii, Ophiothrix fragilis, Ophicoma nigra, Adamsia palliata, Sertularia abietina, Antennularia ramosa, Hydrallmania falcata, Tubularia*, sp., *Glycera*, sp., *Amphiporus pulcher, Flustra foliacea.* It ought to be remembered, however, that a good many (by no means all) of the Mollusca and a few of the Polyzoa were dead.

———

Mr. Clement Reid, F.G.S., of the Geological Survey, has examined the samples of deposits which were sent by us to the Jermyn Street Museum, and reports as follows:—

" The series of dredgings examined since the last report is most interesting from a geological point of view. One is again struck by the common occurrence of loose angular stones at places and depths apparently well beyond the reach of any bottom drift—at least beyond the reach of currents likely to move such coarse material. This stony sea-bed is in all probability the result of submarine erosion of glacial deposits. Its occurrence renders comparison between recent marine deposits of these latitudes and Tertiary deposits a task of peculiar difficulty; for not only is the nature of the true marine sediments masked, but the fauna also must be greatly altered. It is evident that numerous species which need a firm base on which to affix themselves will be encouraged by a stony bottom; while in a Tertiary deposit, formed under identical conditions, except for the absence of stones, they may be entirely missing, having nothing but dead shells to which to attach themselves.

" Notwithstanding this peculiarity of most of the dredgings, a few samples may well be compared with our Older Pliocene (Coralline Crag). I would particularly draw attention to certain localities where material almost entirely of organic origin has been obtained. Of these perhaps the most interesting are some samples full of *Cellaria fistulosa* (found to the south-east of the Calf Sound, 20 fathoms). They are in many respects strikingly like certain parts of the Coralline Crag. The more ordinary type of Coralline Crag, with its extremely varied polyzoon fauna, we cannot yet match in British seas :* it was probably formed, as the mollusca indicate, in a sea several degrees warmer than ours.

" It was hoped that in the course of these dredgings some light might be thrown on the Tertiary strata under-lying the bed of the Irish Sea, for in the North Sea the dredge occasionally brings up hauls of Tertiary fossils. This expectation has not yet been realised ; but possibly, by dredging in the channels where the submarine scour is greatest, such deposits may yet be reached. It is very important to obtain some knowledge of the Tertiary bed of the Irish Sea, for Irish Pleistocene deposits contain a considerable admixture of extinct forms, which may be derived from Tertiary deposits below the sea-level. The Glacial Drift of Aberdeenshire contains Pliocene Volutes and Astartes, derived from some submarine deposit off the Aberdeenshire coast. The so-called ' Middle Glacial Sands ' of Norfolk are full of shells which I now believe to be derived from some older deposit, probably beneath the sea."

* See, however, the deposit described on p. 64, where nearly 60 species of Polyzoa are recorded from one haul.—W. A. H.

The important influence of the shore rocks upon the littoral fauna has not been neglected, and lists and observations are accumulating, but that subject must be left over for a fuller discussion in next year's report.

OTHER INVESTIGATIONS.

Several new lines of investigation have been started during the year, and are still in progress. One of these may be called the "larval-attachment inquiry," and consists in sinking in various parts of the bay an apparatus composed of a rope weighted at one end and buoyed at the other, and having a number of slips of glass, slate, wood, &c., attached at equal distances along its length. These ropes are hauled up and examined periodically, and may be expected, when further observations have been taken, to give information as to the times and modes of attachment of the larvæ of various species, and also as to the most suitable substances for particular kinds of larvæ to settle down upon. So far, glass seemed in the early spring (February and March) to be the favourite substance. A surprisingly large number of algæ, compared with the animals, appeared, and nearly all were on the glass slips. Later on, in the summer, Barnacles (*Balanus*) made their appearance in great numbers on the slips of wood and on the wooden buoy at the top of the apparatus, while all the upper part of the rope within a few feet of the surface became covered with algæ. A number of Ascidians (*Ascidiella virginea*) were also found, in August, to have attached themselves to the rope, and seemed to have got as far as possible in between the strands and into the coils of the knots. On the upper pieces of slate, and in one instance on a piece of glass, there were young specimens of the tubicolous Annelid, *Pomatoceros triqueter*, in no case more than $\frac{1}{2}$ to $\frac{3}{4}$ inch in length.

At the end of October another rope, which had been sunk in the bay since June with a bag of oysters, was hauled up, and the upper 4 or 5 feet, much covered with algæ, was removed for examination in the laboratory. It was found to have the following organisms adhering:—

ALGÆ (identified by Prof. Harvey Gibson):

Ceramium rubrum, C. Ag.

C. strictum, Harv.

C. deslongchampsii, Chauv.

Chantransia daviesii, Thur.

C. virgatula, Thur.

Desmarestia aculeata, Lamx.

Polysiphonia urceolata, Grev.

P. nigrescens, Grev.

P. elongata, Grev.

Dictyota dichotoma, Lamx.

Sphacelaria cirrhosa, C. Ag.

Enteromorpha clathrata, J.Ag.

Monostroma witrockii, Born.

Colonies of *Gomphonema*.

ZOOPHYTES (identified by Miss L. R. Thornely):

Obelia geniculata

O. longissima

Cytia johnstoni

Bougainvillea muscus

Opercularella lacerata

POLYZOA (Miss Thornely):

Membranipora pilosa

Eucratea chelata

Scrupocellaria reptans

Schizoporella hyalina

TUNICATA (Miss J. H. Willmer):

Diplosoma gelatinosum

Ascidiella virginea

CRUSTACEA (identified by Mr. A. O. Walker):

Hippolyte varians

Idotea marina

Hyale nilssonii

Apherusa bispinosa

Dexamine spinosa

Gammarus locusta

Lilljeborgia kinahani

Amphithoe rubricata

Podocerus falcatus

Caprella acanthifera.

THE DRIFT BOTTLES AND SURFACE CURRENTS.

In last year's report the scheme for the distribution of drift bottles over the Irish Sea, for the purpose in helping to determine the set of the chief currents, tidal or otherwise, which might influence the movements of fish food and fish embryos, was fully explained. Since then the work has been going on actively, and now at the end of about twelve months one thousand drift bottles in all have been set free. Many have been let out at intervals of ten minutes, or quarter of an hour, or twenty minutes (corresponding to distances of from 3 to 6 miles apart) from the Isle of Man boats when crossing between Liverpool and Douglas—a very convenient line of 75 miles across the middle of the widest part of our area, traversing the " head of the tide " or meeting place of the tidal currents entering by St. George's Channel and the North Channel. Others have been let off from Mr. Alfred Holt's steamers, in going round from Liverpool to Holyhead and in coming down from Greenock. Mr. Dawson on the Fisheries steamer " John Fell " has distributed a number along the coast in various parts of the district, and the Fisheries bailiffs have let off some dozens from their small boats. Other series have been set free at stated intervals during the rise and fall of the tide from the Morecambe Bay Light Vessel in the northern part of our area, north of the " head of the tide;" and, through the kindness of Lieutenant M. Sweny, R.N., a similar periodic distribution has taken place from the Liverpool North-West Light Vessel, to the south of the " head of the tide." Others, finally, have been despatched by Mr. R. L. Ascroft, by Mr. Andrew Scott and by various members of the Committee in other parts of the area from small boats and on our dredging expeditions, in some cases between the Isle of Man and Ireland. Altogether we have pretty well covered this northern area of the Irish Sea in our distribution of floating bottles.

The first bottles, and the printed paper they contained, were described last year. We afterwards adopted a rather larger size of bottle, 8·5 cm. in length ; and, after various postal difficulties and experiments, we hit upon a convenient size and thickness of private post card, which, ready stamped and addressed, and marked with a distinguishing letter or number, is rolled up in its bottle and has printed on its back the following :—

For scientific enquiry into the currents of the Sea.

Whoever finds this is earnestly requested to write distinctly the DATE and LOCALITY, with full particulars, in the space below, and to put the card in the nearest post office.

[No. here]

LOCALITY, where found..............................

....

....

DATE, when found..............................

Name and address of sender..........................

....

[No. here]

The number is marked with blue and black pencils in duplicate on opposite corners of the card, in case of one edge of the card getting worn by moisture; and the card is so rolled in the bottle that one of these numbers can be read through the glass, in order that a record may be kept of when and where each bottle is set free. Mr. Andrew Scott, Fisheries Assistant at University College, has collated these records with the particulars of finding of those bottles which have been recovered, and I am indebted to him for the details upon which the following statement of results is based.

Altogether out of the 1000 bottles distributed, over 340 or 34 per cent., more than one in three, have been subsequently picked up on the shore, and the paper or post card has been duly filled up and returned to me. I beg to thank the various finders of the bottles for their kindness in filling and posting the cards. They come from various parts of the coast of the Irish Sea—Scotland, England, Wales, Isle of Man, and Ireland. Some of the bottles have gone quite a short distance, having evidently been taken straight ashore by the rising tide; while others have been blown ashore by the wind, *e.g.*, two (post cards 211 and 214) let off near New Brighton stage on 9th October, 1895, the tide ebbing and the wind N.N.W., were found next day near the Red Noses, 1 mile to the west. Others have been carried an unexpected length, *e.g.*, one (No. 35), set free near the Crosby Light Vessel, off Liverpool, at 12.30 p.m., on October 1st, was picked up at Saltcoats, in Ayrshire, on November 7th, having travelled a distance of at least 180 miles* in thirty-seven days; another (H. 20) was set free near the Skerries, Anglesey, on October 6th, and was picked up one mile north of Ardrossan, on November 7th, having travelled 150 miles in thirty-one days; and bottle No. 1, set free at the Liverpool Bar on September 30th, was picked up at Shiskin, Arran, about 165 miles off, on November 12th. On the other hand, a bottle (J. F. 34) set free on November 7th, in the Ribble Estuary, was picked up on November 12th at St. Anne's, having only gone 4 miles.

It may be doubted whether our numbers are sufficiently large to enable us to draw very definite conclusions. It is only by the evidence of large numbers that the vitiating effect of exceptional circumstances, such as an unusual gale, can be eliminated. Prevailing winds, on the other hand, such as would usually affect the drift of surface organisms, are amongst the normally acting causes which we are trying to ascertain. Mr. W. E. Plummer of the Bidston Observatory has kindly given us access to his

---

* More probably, very much further, as during that time it would certainly be carried backwards and forwards by the tide.

records of weather for the last twelve months, and we have noted opposite the bottles, from whose travels we are drawing any conclusions, an approximate estimate of the wind influences during the period when the bottle may have been at sea. There have been a few rather extraordinary journeys, e.g., one let off in the middle of Port Erin Bay on April 23rd was found at Fleetwood on July 6th; another let off at Bradda Head on June 3rd was found on Pilling Sands (near Fleetwood) on July 24th.

It is important to notice that the bottles may support one another's evidence, those set free about the same spot often being found in the same locality, e.g., out of a batch of 6 set free off New Brighton, on Oct. 9th, 1895, 5 have come back and all were found at about the same place.

Dr. Fulton, who is conducting a similar inquiry by means of drift bottles, in the North Sea, for the Scottish Fishery Board, writes to me that he is now having large numbers of his bottles returned to him from the Continent, chiefly Schleswig and Jutland. And he draws the conclusion, "There is no doubt that the current goes across, down as far as Norfolk—none of the bottles have been found south of Lincoln and none in Holland—and this will explain the presence of banks and shallows in the south and east and the immense nurseries of immature fish there." No detailed account of these experiments on the Scottish coast has yet appeared, and it will be interesting to compare our results with them at some future period.*

What is already well-known† in regard to the tidal

---

* Since the above was in type an account has been published.

† All the accounts I have had access to seem based upon Admiral Beechey's observations published in the Philosophical Trans. for 1848 and 1851. Admiral Wharton, F.R.S., the present Hydrographer to the Navy, has kindly informed me that Admiral Beechey took his observations by the direct method of anchoring his ship in various places and then observing the direction and force of the tide.

streams or currents in the Irish Sea is that for nearly six hours after low-water at, say, Liverpool, two tidal streams pour into the Irish Sea, the one from the north of Ireland, through the North Channel, and the other from the southward, through St. George's Channel. Parts of the two streams meet and neutralise each other to the west of the Isle of Man, causing the large elliptical area, about 20 miles in diameter and reaching from off Port Erin to Carlingford, where no tidal streams exist, the level of the water merely rising and falling with the tide. The remaining portions of the two tidal streams pass to the east of the Isle of Man and eventually meet along a line extending from Maughold Head into Morecambe Bay. This line is the " head of the tide." During the ebb the above currents are practically reversed, but in running out the southern current is found to bear more over towards the Irish coast.

There is some reason to believe that, as a result of the general drift of the surface waters of the Atlantic and the shape and direction of the openings to the Irish Sea, more water passes out by the North Channel than enters that way, and more water enters by the South (St. George's) Channel than passes back, and that consequently there is, irrespective of the tides, a slow current passing from south to north through our district. The fact that so many of our drift bottles have crossed the " head of the tide " from S. to N., and that of those which have gone out of our district nearly all have gone north to the Clyde Sea-area supports this view, which I learn from Admiral Wharton is *a priori* probable, and which is believed in by some nautical men in the district from their experience of the drift of wreckage.

It may be objected to our observations by means of drift bottles that they are largely influenced by the wind and waves, and are not carried entirely by tidal streams. Well, that is an advantage rather than any objection to the method. For our object is to determine not the tidal currents alone but the resulting effect upon small surface organisms such as floating fish eggs, embryos and fish food, of *all* the factors which can influence their movements, including prevalent winds. The only factors which can vitiate our conclusions are unusual gales or any other quite exceptional occurrences, and the only way to eliminate such influences is (1) to allow for them so far as they are known from the weather reports, and (2) to employ a large number of drift bottles and continue the observations over a considerable time. We have carefully considered the bearing of the weather records, and we think that the large number of bottles we have made use of, during the year, ought to enable us to come to some definite results. Our conclusions so far (November) then are :—

(1) A large number (over 34 %) have been stranded and found and returned, (2) only a small proportion (13 %) have been carried out of our part of the Irish Sea, (3) nearly 12 % have crossed the "head of the tide" showing the influence of wind in carrying floating bodies over from one tidal system into another, (4) most of the bottles set free to the west of the Isle of Man have been carried across to Ireland, only a small number (3·8 %) of them have got round to the eastern side of the Island and been carried ashore on the English coasts, (5) the majority of the bottles set free off Dalby have gone to the Co. Down coast, (6) a considerable number of bottles have been set free over the deep water to the east of the Isle of Man, where our more valuable flat fish spawn, and of those that have been returned the majority had been carried to the Lancashire, Cheshire and Cumberland coasts. So we may reasonably conclude that the embryos

of fish spawning off Dalby would tend to be carried across to the Irish Coast, while those of fish spawning in the deep water on this eastern side of the Isle of Man would go to supply the nurseries in the shallow Lancashire and Cheshire bays, and very few would be carried altogether out of the district.

OTHER FAUNISTIC WORK.

Mr. Arnold Watson writes as follows in regard to the Annelids at which he is working :—" The most interesting item is probably the capture by Mr. R. L. Ascroft, in May last year, and subsequently at intervals, of larval *Pectinaria* in the waters near Blackpool. The specimens were taken with a tow-net attached to the beam of a trawl, and show that in the first instance the animal secretes a minute free tube of organic matter of somewhat cellular appearance. This tube is about $\frac{1}{23}$ of an inch long, $\frac{1}{100}$ of an inch in its widest diameter, tapering to about $\frac{1}{200}$ at the narrow end. To the wider end of this membranous tube the worm attaches very minute grains of sand, course after course, forming a sand tube about $\frac{1}{90}$ of an inch in external diameter. The length of the larval worm from tip of tail to the outer margin of the minute headbristles, or combs, is about $\frac{1}{25}$ of an inch. Last spring Mr. Ascroft was good enough to send me some living specimens of these larvæ which, for a few days, survived their journey, and were very active. At this stage of the animal's existence a pair of eye-spots are visible. He also sent me in March last, for identification, a specimen of *Autolytus alexandri* (with its egg sac) taken by surface tow-netting in the daytime off the Bahama Light Ship, near Ramsey, Isle of Man. Hornell recorded in 1892 a MALE specimen of this worm taken by tow-netting off Puffin Island, which was the first recorded from British waters."

Mr. Watson has completed his work on the tube-building habits of *Panthalis oerstedi*, referred to in last report, and his paper on the subject, with two plates, has been published in Vol. IV. of the "Fauna." One of the specimens of *Panthalis* from Port Erin lived in Mr. Watson's Aquarium at Sheffield from September 30th, 1894, to October 8th, 1895, when as it seemed ailing he killed it with corrosive.

Dr. J. D. F. Gilchrist who has paid several short visits to the Biological Station, and worked there for some time at Easter, has sent in the following report upon his work to the Director :—" During my stay at the Marine Station at Port Erin, I was chiefly concerned with the Mollusca, but found the frequent shore collecting and dredging excursions very profitable for general work. *Aplysia* was found in abundance by dredging and I took this opportunity of trying various methods of killing the animal in an expanded condition. After trying several, I found the following the only method which could be depended on with certainty. A few drops of a 5 % solution of cocain were mixed with the water in which the Aplysias were. After a time they expanded fully. They were then left in the solution (12 hours or more) till no contraction took place when removed and put into weak alcohol. If contraction took place they would be put back into the cocain solution when they again expanded. This was repeated till no contraction took place, when they could, after a time, be put into stronger alcohol. Other methods though simpler, and not so tedious, were less dependable and at best gave a somewhat abnormally inflated appearance.

" At Prof. Herdman's suggestion a solution of formol was tried as a preserving fluid for *Aplysia* and *Pleurobranchus*. In both cases a considerable amount of colouring matter was dissolved out of the integument and stained the surrounding fluid of a reddish colour.

" A series of experiments on the method of feeding in Lamellibranchs was begun, to show in what manner the gills exercised the function of collecting food material and the labial palps of discriminative selection of food particles. I hope to be able to give the results of these experiments after further observations.

" I also procured at the Station specimens of Opisthobranchs which will form material for future work."

The Rev. T. S. Lea (who has kindly presented a large ordinance map, 6 inches to the mile, of the S.W. of the Isle of Man, to the Biological Station) has continued this summer his series of observations upon the zones of algæ on the shore, and has taken a number of photographs of species *in situ* on the rocks and in pools.

The work done by Mr. Browne and by Mr. Beaumont at the Station is sufficiently dealt with in their reports; upon the Medusæ and the Nemertea respectively, which have now been published in Vol. IV. of the " Fauna." Mr. Walker and Mr. Thompson have discussed the results of their work on Crustacea some pages back (p. 42); and the investigations of Prof. Boyce and Prof. Herdman on oysters under various normal and abnormal conditions is work of a very special nature—partly bacteriological and partly experimental—which is still in progress, and will be reported upon in full to the British Association and to the Lancashire Sea-Fisheries Committee.

In a brief report from the Curator giving an outline of the work of the summer the following observations which seem worthy of permanent record occur:—" The dredgings in May produced amongst other nudibranchs *Cuthona arenicola*, new to the fauna round the Isle of Man ; and two Polychætes, which though common here have not, as yet, been recorded in the L. M. B. C. lists, were found on the shore—these were *Amphitrite johnstoni* and *Arenicola ecaudata*. The latter species seems to take the place of *A. marina* amongst stones and muddy shingle where it is invariably found, while *A. marina* is confined to the sand...The dredgings in June brought to light several interesting animals, some of the more important finds being:—*Cratena olivacea*, new to the Isle of Man, and which has proved to be not uncommon in the upper Coralline zone off Port Erin; *Embletonia pulchra*, new to the district, which during June and July appeared in almost every dredging that was taken; *Coryphella landsburgi*, new to the Isle of Man, taken several times; *Oscanius membranaceus*, dredged in 15 fathoms off Port Erin; and *Eolis concinna*. At Whitsuntide, *Polygordius* was dredged from a gravel bottom off Bradda Head, and it has since turned up in two different localities, from similar ground... About this time of the year (July) and for the rest of the summer the bay was full of young fish known to the fishermen as "Gilpins." These are chiefly young cod and pollach—mostly the latter. Some were caught and put into one the tanks, several are still alive and have grown considerably since their capture...Towards the end of the month the annual cleaning of the buoy took place. This year there were several tubes and worms of what is probably *Sabella penicillus*...There were, as before, great quantities of Caprellids, *Ascidiella virginea* and *Ciona intestinalis* were present, along with the nudibranchs *Facelina drummondi*, *Cuthona aurantiaca*, and *Dendronotus arborescens*...During the spring months, until towards the middle of June, *Aplysia punctata* was one of the commonest animals in the bay. It came up in quantities in the dredge, it was to be found commonly on the shore amongst rocks, and after a westerly wind the shore was covered with masses of its spawn. After June *Aplysia* almost entirely disappeared, and was not found again until the end of September...The following common animals could usually be supplied alive to laboratories and museums at any time without much delay :—*Actinia mesembryanthemum, Tealia crassicornis, Bunodes gemmaceus, Actinoloba dianthus, Alcyonium digitatum, Echinus esculentus, E. miliaris, Ophiothrix fragilis, Ophiocoma nigra, Ophiura ciliaris, Arenicola marina* and *A. ecaudata, Nereis pelagica, Pecten maximus* and *P. opercularis, Doris tuberculata, Aplysia punctata* (spring), *Cancer pagurus, Carcinus moenas, Nephrops norvegicus, Galathea squamifera.*"

There are of course many other animals, both common species and rarer forms, which could be obtained alive by giving a little notice, or preserved in spirit, on applying to the Hon. Director, at University College, Liverpool.

The little reference library at the Station is growing gradually, but is still badly in want of many common books and pamphlets. Any works on Marine Zoology, on British Animals, or on the structure and development of marine invertebrates will be thankfully received. The Committee are much indebted to Prof. G. B. Howes who has kindly presented to the Station a series of 7 volumes of collected fishery papers—the result of the Fisheries Exhibition of 1883. Other books and pamphlets have been received from members of the Committee, and the following books have been purchased during the year :— Baird's British Entomostraca, Johnston's British Museum Catalogue of Worms, and Jeffrey's British Conchology (5 vols.).

PUBLICATIONS.

Since the last Annual Report, we have issued Vol. IV. of the "Fauna of Liverpool Bay." It is the largest volume of the series and contains about 475 pages and 53 plates, several of which are coloured. In addition to the reports and papers which had been already announced as forthcoming in this volume, Vol. IV. contains a note upon the yellow variety of *Sarcodictyon* by Prof. Herdman, a paper on the structure of the cerata of *Dendronotus* by Mr. J. A. Clubb, a revision of the Amphipoda of the L. M. B. C. District by Mr. Walker, and a supplementary report on the Port Erin Nemertines by Mr. Beaumont.

It ought to be noticed that although the primary objects of the Committee were originally faunistic and speciographic, yet observations on habits and life-histories, and bionomics in general, have not been neglected; and now some of our papers in this Vol. IV., such as Mr. Chadwick's on the Vascular Systems of the Starfishes, and Mr. Clubb's on the Cerata of Nudibranchs, are coming to deal with purely structural and morphological questions.

The other Reports in this volume deal, some of them— such as Mr. Gamble's on Turbellaria, Mr. Beaumont's on Nemertea, and Mr. Browne's on Medusæ—with fresh groups of animals which had not been adequately discussed in the previous volumes ; while others, such as Mr. Thompson's and Mr. Walker's reports, are welcome revisions of these authors' own previous work on the Crustacea. Dr. Hanitsch has furnished us with a paper on the Classification and Nomenclature of British Sponges, which it may be said does not come strictly within the scope of the L.M.B.C. Reports. Still the subject matter is of such importance to anyone working systematically at our sponge fauna, and the treatment seems so well adapted to render the lists an indispensable working addition to Bowerbank's Monograph, that I had no hesitation in asking Dr. Hanitsch to allow the paper to be included in our series of reports upon the Fauna of Liverpool Bay.

As to the future, there are a number of reports upon groups, and other pieces of work, in progress. The " List of Fishes " is still in hand.   Mr. Andrew Scott has undertaken to collect and report upon the Ostracoda, Dr. Hurst has still charge of our Pycnogonida ; while Prof. Boyce and Prof. Herdman are engaged on an extensive investigation on Oysters in healthy and in diseased conditions which has been partly laid before the British Association, but ought to be published in full next year after some further series of observations and experiments have been made.

The Infusoria of all kinds, some of the parasitic groups of Crustacea, the marine Rotifera, and some of the lower worms are still not allotted to workers ; while there is plenty for many hands to do in working out the detailed distribution of genera and species, and in tabulating and discussing the results' of dredging in various depths and localities.

There is no need to dwell upon the large number of species now recorded, and the additions that have been made by our explorations both to the British fauna and to science ; such results, though very necessary, are no longer the sole, perhaps not even the chief objects which the Committee have in view.   I think all who are engaged in this L. M. B. C. work feel that it is growing steadily under their hands in every direction.. Not only are there many animals and whole groups of animals in our sea awaiting examination and record, but there are many points of view, the speciographic, distributional, anatomical, physiological, embryological, bionomical and others, from which even the best known forms would well repay further and more detailed investigation ; and wider problems such as the association of animals together on particular sea-bottoms and at particular depths, and other questions of bionomics and of oceanography—some of them having important bearings upon Geology and upon Fishery questions—are now opening up before us and pressing for solution.

We are a small body, the Naturalists of Liverpool, our laboratory at Port Erin is a modest establishment with but scanty equipment, we have no State, County or Municipal subsidies, and our available funds (private subscriptions) are barely sufficient for the necessary expenses of steamer and apparatus in our explorations, and for the publication of our results; but fortunately there is no lack of work for us to do, work which is interesting in the doing, and work which, if we seek it earnestly and do it honestly, we cannot but believe will be of value to science, and may, through its connection with the fishing industries, be of direct benefit to mankind.

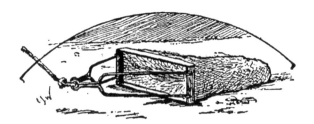

APPENDIX A.

LIVERPOOL MARINE BIOLOGICAL STATION at PORT ERIN.

REGULATIONS.

I.—This Biological Station is under the control of the Liverpool Marine Biology Committee, the executive of which consists of the Hon. Director (Prof. Herdman, F.R.S.) and the Hon. Treasurer (Mr. I. C. Thompson, F.L.S.).

II.—In the absence of the Director, and of all other members of the Committee, the Station is under the temporary control of the Resident Curator or Laboratory Assistant, who will keep the keys, and will decide, in the event of any difficulty, which places are to be occupied by workers, and how the tanks, collecting apparatus, &c., are to be employed.

III.—The Resident Assistant will be ready at all reasonable hours and within reasonable limits to give assistance to workers at the Station, and to do his best to supply them with material for their investigations.

IV.—Visitors will be admitted, on payment of a small specified charge, to see the Aquarium and the Station, so long as it is found not to interfere with the scientific work.

V.—Those who are entitled to work in the Station, when there is room, and after formal application to the Director, are :—(1) Annual subscribers of one guinea or upwards to the funds (each guinea subscribed entitling to the use of a work place for four weeks), and (2) others who are not annual subscribers, but who pay the Treasurer

10s. per week for the accommodation and privileges. Institutions, such as Colleges and Museums, may become subscribers in order that a work place may be at the disposal of their staff for a certain period annually: a subscription of two guineas will secure a work place for six weeks in the year, a subscription of five guineas for four months, and a subscription of £10 for the whole year.

VI.—Workers at the Station can always find comfortable and convenient quarters at the closely adjacent Bellevue Hotel; but lodgings can readily be had by those who prefer them.

VII.—Each worker is entitled to a work place opposite a window in the Laboratory, and may make use of the microscopes, reagents, and other apparatus, and of the boats, dredges, tow-nets, &c., so far as is compatible with the claims of other workers and with the routine work of the Station.

VIII.—Each worker will be allowed to use one pint of methylated spirit per week, free. Any further amount required must be paid for. All dishes, jars, bottles, tubes, and other glass may be used freely, but must not be taken away from the laboratory. If any workers desire to make, preserve, and take away collections of marine animals and plants, they must make special arrangements with the Director or Treasurer in regard to bottles and preservatives. Although workers in the Station are free to make their own collections at Port Erin, it must be clearly understood that (as in other Biological Stations) no specimens must be taken for such purposes from the laboratory stock, nor from the Aquarium tanks, nor from the steam-boat dredging expeditions, as these specimens are the property of the Committee. The specimens in the Laboratory stock are preserved for sale, the animals in the tanks are for the instruction of visitors to the

Aquarium, and as all the expenses of steam-boat dredging expeditions are defrayed by the Committee the specimens obtained on these occasions must be retained by the Committee (*a*) for the use of the specialists working at the Fauna of Liverpool Bay, (*b*) to replenish the tanks, and (*c*) to add to the stock of duplicate animals for sale from the Laboratory.

IX.—Each worker at the Station is expected to lay a paper on some of his results—or at least a short report upon his work—before the Biological Society of Liverpool during the current or the following session.

X.—All subscriptions, payments, and other communications relating to finance, should be sent to the Hon. Treasurer, Mr. I. C. Thompson, F.L.S., 19, Waverley Road, Liverpool. Applications for permission to work at the Station, or for specimens, or any communications in regard to the scientific work should be made to Professor Herdman, F.R.S., University College, Liverpool.

APPENDIX B.

SUBSCRIPTIONS and DONATIONS.

	Subscriptions.			Donations.		
	£	s.	d.	£	s.	d.
Ayre, John W., Ripponden, Halifax ...	1	1	0	—		
Banks, Prof. W. Mitchell, 28, Rodney-st.	2	2	0	—		
Beaumont, W. I., Cambridge	2	2	0	—		
Bickersteth, Dr., 2, Rodney-st.	2	2	0	—		
Boulnois, H. P., 7, Devonshire-rd. ...	1	1	0	—		
Brown, Prof. J. Campbell, Univ. Coll. ...	1	1	0	—		
Browne, Edward T., B.A., 141, Uxbridge-						
road, Shepherd's Bush, London ...	1	1	0	—		
Boyce, Prof., University College ...	1	1	0	—		
Caton, Dr., 31, Rodney-street	—			1	1	0
Clague, Dr., Castletown, Isle of Man ...	1	1	0	—		
Clague, Thomas, Bellevue Hotel, Port Erin	1	1	0	—		
Comber, Thomas, J.P., Leighton, Parkgate	1	1	0	—		
Crellin, John C., J.P., Ballachurry, An-						
dreas, Isle of Man	0	10	6	—		
Darbishire, R.D., Victoria-pk., Manchester	1	1	0	4	4	0
Dawkins, Professor W. Boyd, Owens						
College, Manchester...	1	1	0	—		
Derby, Earl of, Knowsley	5	0	0	—		
Delius, W. Meyer, Hamburg	2	2	0	—		
Dumergue, A. F., 7 Montpellier-terrace	0	10	6	—		
Gair, H. W., Smithdown-rd., Wavertree	2	2	0	—		
Gamble, Col. C.B., Windlehurst, St. Helens	2	0	0	—		
Gamble, F.W., Owens College, Manchester	1	1	0	—		
Gaskell, Frank, Woolton Wood... ...	1	1	0	—		
Gaskell, Holbrook, J.P., Woolton Wood	1	1	0	—		
Gell, James S., High Bailiff of Castletown	1	1	0	—		
Gibson, Prof. R. J. H., 41, Sydenham-av.	1	1	0	—		

Gifford, J., Whitehouse-terrace, Edin. ...	1	0	0	—		
Gilchrist, Dr. J. D. F., Edinburgh Univ.	1	1	0	—		
Glynn, Dr., 62, Rodney-street	2	2	0	—		
Greening, Linnæus, 5, Wilson Patten-st., Warrington	1	1	0	—		
Gotch, Prof., Museum, Oxford	1	1	0	—		
Halls, W. J., 35, Lord-street	1	1	0	—		
Henderson, W. G., Liverpool Union Bank	1	1	0	—		
Herdman, Prof., University College ...	2	2	0	—		
Holder, Thos., 1, Clarendon-buildings, Tithebarn-street	1	1	0	—		
Holland, Walter, Mossley Hill-road ...	2	2	0	—		
Holt, Alfred, Crofton, Aigburth	2	2	0	—		
Holt, George, J.P., Sudley, Mossley Hill	1	0	0	—		
Howes, Prof. G. B., Royal College of Science, South Kensington, London	1	1	0	—		
Hoyle, W. E., Museum, Owens College, Manchester	1	1	0	—		
Isle of Man Natural History and Anti- quarian Society	1	1	0	—		
Jones, C.W.,J.P., Field House,Wavertree	1	0	0	—		
Kermode, P. M. C., Hill-side, Ramsey...	1	1	0	—		
Lea, Rev. T. Simcox, 3, Wellington-fields	1	1	0	—		
Leicester, Alfred, Harlow, Essex ...	1	1	0	—		
Liverpool Museum Committee	2	2	0	—		
Macfie, Robert, Airds	1	0	0	—		
Meyer, Dr. Kuno, University College ...	0	5	0	—		
Meade-King, H. W., J.P., Sandfield Park	1	1	0	—		
Meade-King, R. R., 4, Oldhall-street ...	0	10	0	—		
Melly, W. R., 90, Chatham-street ...	1	1	0	—		
Miall, Prof., Yorkshire College, Leeds ...	1	1	0	—		
Michael, Albert D., Cadogan Mansions, Sloane Square, London, S.W. ...				1	1	0
Monks, F. W., Brooklands, Warrington	1	1	0	—		
Muspratt, E. K., Seaforth Hall	5	0	0	—		
Newton, John, M.R.C.S., 44, Rodney-st.	0	10	6	—		

Poole, Sir James, Tower Buildings ...	2	2	0	—		
Rathbone, S.G., Croxteth-drive, Sefton-pk.	2	2	0	—		
Rathbone, Mrs. Theo., Backwood, Neston	1	1	0	—		
Rathbone, Miss May, Backwood, Neston	1	1	0	—		
Rathbone, W., Greenbank, Allerton ...	2	2	0	—		
Roberts, Isaac, F.R.S., Crowborough ...	1	1	0	—		
Shaw, Prof. H. S. Hele, Ullet-road ...	1	1	0	—		
Shepheard, T., Kingsley Lodge, Chester	1	1	0	—		
Simpson, J. Hope, Annandale, Aigburth-dr	2	2	0	—		
Smith, A. T., junr., 24, King-street ...	1	1	0	—		
Talbot, Rev. T. U., 4, Osborne-terrace,						
Douglas, Isle of Man	1	1	0	—		
Thompson, Isaac C., 19, Waverley-road	2	2	0	—		
Thornely, James, Baycliff, Woolton ...	1	1	0	—		
Thornely, The Misses, Baycliff, Woolton	1	1	0	—		
Toll, J. M., 340, Walton Breck-road ...	1	1	0	—		
Turnbull, Thos. S., 18, Spring-gardens,						
Manchester	1	1	0	—		
Walker, A. O., Nant-y-glyn, Colwyn Bay	3	3	0	—		
Walker, Horace, South Lodge, Princes-pk.	1	1	0	—		
Walters, Rev. Frank, B.A., King William						
College, Isle of Man...	1	1	0	—		
Watson, A. T., Tapton-cresent, Sheffield	1	1	0	—		
Weiss, Prof. F. E., Owen's College, Man'tr.	1	1	0	—		
Westminster, Duke of, Eaton Hall ...	5	0	0	—		
White, Prof., University College, Bangor	2	0	0	—		
Wiglesworth, Dr., Rainhill	1	1	0	—		
	109	6	6	6	6	0

THE LIVERPOOL MARINE BIOLOGY COMMITTEE.

In Account with ISAAC C. THOMPSON, Hon. Treasurer.

Dr.

1895.	£	s.	d.
To Printing Reports, Plates, &c.,	20	17	9
,, Printing and Stationery	1	12	8
,, Expenses of Dredging Expeditions	28	15	7
,, Boat Hire	0	7	6
,, Books and Apparatus at Port Erin Biological Station	10	13	3
,, Postage, Carriage of Specimens, &c.,	3	16	11
,, Salaries, Curator and Laboratory Boy	50	3	4
,, Rent of Port Erin Biological Station	15	0	0
,, Repairs, ,, ,, ,,	3	17	2
,, Sundries	0	10	5
	£135	14	7

Cr.

1895.	£	s.	d.
By Balance in hand, Dec. 31st, 1894	10	16	2
,, Subscriptions and Donations actually received	105	8	6
,, Dividend, British Workman's Public House Co., Ltd., Shares	5	18	9
,, Sale of Reports	4	13	6
,, Bank Interest	0	19	9
,, Admissions to Aquarium	2	4	6
,, Balance due Treasurer	5	13	5
	£135	14	7

Endowment Investment Fund :—
British Workmans' Public House Co's. Shares   ...£173 1 0

Audited and found correct,

A. T. SMITH, Junr.

ISAAC C. THOMPSON,

Hon. Treasurer.

LIVERPOOL, *December 31st*, 1895.

FREE-SWIMMING COPEPODA from the WEST COAST of IRELAND.

By Isaac C. Thompson, F.L.S.

[Read January 10th, 1896.]

A NUMBER of small bottles (eighteen bottles in all) containing tow-net material have been recently handed to me for examination and identification by my friend Mr. Edward T. Browne, B.A., of London, he having collected them off Valencia on the West of Ireland during the summer and autumn of 1895.

They represent the results of sixteen separate days collecting, the dates being April 5, 8, 10, 12, 13, 15, 16, 18, 27, 29, May 5, 8, 14, 27, June 27, July 8, Sept. 6, and Oct. 16, and these corresponding to gatherings Nos. 1 to 18 consecutively (see Table I., at end).

Mr. Browne writes "I did not preserve every tow-netting taken, or keep the whole, but only a sample of it." The sizes of the bottles varied from 2 drms. up to 2 oz., the preservative material used being in most cases a 5 % solution of formalin, the merits of which I shall have occasion to refer to later on. It is to be regretted that "only a sample" of the haul was retained, as experience has often demonstrated that although the mass of a tow-netting may contain mainly one or a very few species, rarer species may occur isolated throughout it, and the very last dip sometimes contains an unexpected prize. So that although the process is slow and tedious, it is always advisable to examine as much as possible, and the careful observer will generally find himself rewarded by so doing.

The free-swimming Copepoda of our coasts vary in size from about 1 mm. to 4 mm. in length.   The mode of examination which I have found the best and quickest is as follows :—After carefully shaking the material in the bottle a quantity is poured into a shallow open glass plate about 4 inches long, 2 inches wide and ⅛ inch deep.   Such a plate (which I have been able to obtain only from Messrs. Cogit & Co., Paris,) is curved inside like a watch-glass and the contents can be rapidly gone over by means of a strong lens, or a simple dissecting microscope, or still better on the large flat stage of a Swift's Stephenson Binocular Microscope, using a 2 inch objective.   Every portion is thus systematically examined, and those animals identified or required for further examination are easily picked out with a very fine needle, or better with the lower part of a cat's whisker cut flat at the end and mounted.

The material of Mr. Browne's collection may all be classed as "Littoral Plankton" being taken at the surface or at a depth of from 1 to 5 fathoms, one only of the number (No. 5) being taken at a depth of from 5 to 15 fathoms.

As before stated Nos. 1 to 10 were collected during the month of April, Nos. 11 to 14 during May and the others during June, July, September, and October.   The localities are very adjacent to each other, the greater number of the tow-nettings having been taken in Valencia Harbour, and the rest off Beginnis Island and Doulus Bay.   Twenty-two species of Copepoda in all were found. In order to conveniently show their distribution and abundance or scarcity I have tabulated them, the numbers running in the order of dates (Table I.).

One species only, *Calanus finmarchicus* was common to all and it formed nearly the entire bulk of many of the earlier bottles. As the main food of the Greenland whale, this species is very abundant in Arctic Seas and is commonly found around our coasts during the winter and spring, becoming scarce or almost disappearing during the warmth of summer.

The more delicate free-swimmers *Oithona spinifrons* and *Acartia clausii* on the contrary it will be seen both appear in the middle of April (No. 6) and continue generally present in the subsequent tow-nettings throughout the summer. A reference to the distribution table will show that the late autumn gathering of October 16th was much the most prolific in species, fifteen species being then taken.

Six of the species found are decidedly rare, *viz.*, *Metridia armata*, *Candace pectinata*, *Pseudocalanus armatus*, *Monstrilla rigida*, *Corycæus speciosus*, and *Oncæa mediterranea*. The first named, a few specimens occurring in Nos. 11 and 17 and 18, is usually a surface animal and its first recorded British habitats are in the Valencia neighbourhood (Brady's " Free and Semi-parastic Copepoda of the British Islands," Vol. I., p. 42). I have taken it very sparingly in Liverpool Bay and in parts of the Clyde, and Scott reports it from the Forth. *Candace pectinata* occurs in four gatherings, Nos. 5, 14 and 16, and 18. It is usually found at a considerable depth below the surface as is the case with No. 5 specimen; I have found it in tow-nettings taken off the Ross of Mull at 55 fathoms, and about the surface off Oban, and plentifully in the Mediterranean. *Monstrilla rigida*, a single specimen of which occurs in No. 7 has not yet disclosed its life-history. From the absence of functional parts and especially of mouth organs, the species of this mysterious genus seem to give evidence of some other phase of existence, although so far as I am aware no such phase has hitherto been discovered nor have they been recorded as parasitic on any other animal.

A single specimen of *Anomalocera patersonii* occurred in No. 16 bottle. This large and very striking species is very variable in its distribution, often occurring in immense profusion, but is otherwise uncommon. On more than one occasion I have seen the surface of the sea, for many miles around the Isle of Man, so densely covered with this animal as to make it distinctly recognizable from the ship's side, and its beautiful coloration is well known to microscopists.

Pseudocalanus armatus a few specimens of which I found in No. 18 is usually a deep swimmer and never common.

The occurrence of the two southern species *Corycæus speciosus* and *Oncæa mediterranea* is specially interesting as indicating most probably Gulf Stream influence. So far as I am aware the former is new to Britain although it is quite possible that it may have been mistaken for *C. anglicus*, Lubbock, which it strongly resembles. The position of the eyes and the strongly divergent caudal stylets in these specimens seem to clearly indicate its identity with *Corycæus speciosus*, Dana. I found several specimens, some with ovisacs, in No. 18 only, so it appears to have arrived at the end of an unusually hot summer and would probably succumb to the first cold.

Oncæa mediterranea was found sparingly by Mr. G. C. Bourne, M.A., near Plymouth, in 1889 (Report on the Pelagic Copepoda collected at Plymouth in 1888-9), but I am not aware that it has hitherto been reported elsewhere in Great Britain or indeed north of the Mediterranean. I have found it common about the Canary Islands. Two or three specimens only, occurred in No. 18. The other species found are none of them specially noteworthy and are mostly common around our coasts.

In connexion with this collection I would refer naturalists to a paper by Prof. Herdman, F.R.S., entitled "The Biological Results of the Cruise of the S.Y. 'Argo' round the West Coast of Ireland," in August, 1890 (Trans. L'pool Biol. Soc., Vol. V., p. 181). The tow-net material obtained on the "Argo's" cruise was placed in my hands for examination by Prof. Herdman, the results being given in his paper. As a supplement to the present paper I have thought it might be of advantage to other workers on the West Coast to transcribe from Dr. Herdman's paper the tabular statement of the distribution of Copepoda at twelve localities visited by the "Argo." The comparative scarcity of *Calanus finmarchicus* in the "Argo" collection and the prevalence of *Acartia clausii* and *Oithona spinifrons* quite bears out the remarks above made respecting the distribution of those crustaceans. Of the 32 "Argo" species, 15 occur also in the Valencia tow-nettings (see Table II.).

Both collections, but more especially that of Valencia, furnish evidence of the truth of the remarks made by Prof. Herdman in his Presidential Address to the Biological Section at the Ipswich Meeting of the British Association as to the relatively large number of genera of animals represented by the species, in shallow waters, and its bearing on the Darwinian principle that an animal's most potent enemies are its own close allies. In the Valencia group the 22 species recorded belong to 18 genera, the genera being therefore to the species as 9 to 11, and in the "Argo" group 32 species belong to 23 genera or less than 3 to 4. The difference between the two collections in this respect is probably to be accounted for from the fact that the "Argo" collection, besides covering a widely distant area, included a number of ground dwelling, copepoda, or those which are not generally free-swimmers, and would not therefore be brought so much into competition with one another in the struggle for existence as the altogether free-swimmers. Of the latter class may be instanced the genera *Harpacticus*, *Peltidium*, *Laophonte*, and *Cletodes* each of which furnish 2 or 3 species.

I have expressed regret that there was not a larger quantity of material for examination, and have the further regret that we had not dredged deposits from the various localities, for it is to this source mainly that we must look for new species, as has been recently so well exemplified by Messrs. T. and A. Scott in the Clyde area.

As this is a report on the Copepoda only, I need only briefly refer to the other organisms found in the bottles. Most noteworthy was the profusion of *Appendicularia* especially in the spring tow-nettings, where they formed a large percentage of the bulk. The large number of small Medusæ was a conspicuous feature and will no doubt not have escaped the attention of so acute an observer as Mr. Browne himself. Larval Decapoda and *Sagitta* were also abundant throughout.

As before stated a 5 % solution of formaldehyde (formalin) was the preservative used, and the result is all that can be desired and quite confirms my own previous experience of its valuable preservative properties. A drawback to its use when the preserved object is required to be mounted for the microscope is the shrinkage caused to the more delicate forms on removal to Farrants medium or glycerine jelly, though probably this might be overcome by the use of an intermediate solution containing a smaller percentage of glycerine.

In conclusion may I express the hope that while heartily thanking Mr. Browne, he or some other natur-

VALENCIA TOW-NETTINGS—

Date, Locality and Depth (where not stated = surface).	Valencia Hbr. 1½ f. April 5.	do. April 8. 1 f.	do. April 10. 1½ f.	do. April 12. 1½ f.	Donlus Bay, 5-15 f. April 13. 1½ f.	N. side Beginnis Is. 2 f. April 15.	Valencia Hbr. 2 f. April 16.
	1	2	3	4	5	6	7
Calanus finmarchicus ...	×	×	×	×	×	×	×
Pseudocalanus elongatus..		×			×	×	×
P. armatus...							
Centropages hamatus......					×	×	
C. typicus		×			×		
Temora longicornis			×		×	×	
Isias clavipes							
Candace pectinata.........					×		
Metridia armata							
Acartia clausii						×	×
(= Dias longiremis)							
Anomalocera patersonii..							
Parapontella brevicornis.							
Oithona spinifrons						×	×
Thalestris peltata			×				
T. longimana ...							
Laophonte curticauda ...							
Harpacticus chelifer							
Monstrilla rigida							×
Ectinosoma atlanticum...							
E. spinipes							
Corycæus speciosus.........							
Oncæa mediterranea.......							

TABLE I.

DISTRIBUTION OF COPEPODA.

8	9	10	11	12	13	14	15	16	17	18
do. 1-3 f. April 18.	2 f. April 27.	do. April 29.	do. May 5.	do. May 8.	Donlus Bay. May 14	Entrance to Valencia Hbr. May 27.	do. June 27.	do. July 8.	do. Sept. 6	do. Oct. 16.
×	×	×	×	×	×	×	×	×	×	×
×	×	×		×				×	×	×
										×
			×			×	×	×		×
					×		×		×	×
×	×				×	×	×	×	×	×
						×		×		×
			×					×	×	×
×	×	×	×	×	×	×	×	×	×	
								×		
		×								
				×	×	×	×	×	×	×
		×								
										×
								×		×
										×
						×			×	
										×
										×

"ARGO" TOW-NETTINGS—

(From Prof. Herdman's paper "The Biological round the West Coast of Ireland," in August, 1890.

Number in List.	1	2	3, 4
Locality.	Off Pladda.	Lough Swilly.	Gola Id.
SPECIES.			
Calanus finmarchicus	×		
Pseudocalanus elongatus	×	×	
Temora longicornis	×		
Centropages hamatus	×	×	
C. typicus			
Dias longiremis	×		
D. discaudatus			
Isias clavipes			
Pontella wollastoni			
Parapontella brevicornis			
Oithona spinifrons	×	×	
Pseudocyclops obtusus			
Ectinosoma atlanticum	×		
E. erythrops			
E. spinipes			
Longipedia coronata			
Cyclopina littoralis			
Harpacticus chelifer			
do. var. gracilis			
H. fulvus			
Idya furcata			
Thalestris longimana			
Bradya typica			
Peltidium depressum			
P. interruptum			
Laophonte similis			
L. curticauda			
L. longicauda			
Cletodes limicola			
C. linearis			
Monstrilla rigida			
Porcellidium subrotundum			

DISTRIBUTION OF COPEPODA.

Results of the Cruise of the S.Y. "Argo"
Trans. Biol. Soc., L'pool, Vol. V., p. 181.)

5	6, 7	8,9,10	11, 12	13	14	15,16,17	18,19,23	21
Killybegs.	Killary Bay.	Galway Bay.	Killeany Bay, Arran Ids.	Carrigaholt.	Scattery Id.	Templenoe, Kenmare River.	Berehaven.	Glengariff
			X					
X	X	X	X X					X
				X	X	X	X	
X	X		X X X	X X X	X X		X X	X
X			X X	X X	X	X	X	X X
	X			X	X X	X	X	X X X
				X	X	X	X	X X
					X			
					X X X	X X		
	X	X	X	X	X X X	X X	X	X X X
						X		X X X
X X					X			X X
					X			⌣
		⟩			X X X X X	X	X X	
						X	⟩	
				X			⟩	
					X X X			

TABLE II.

alist, may on a future occasion, be able to procure dredged material from the sea bottom and low water mark from the localities referred to in this report, so that we may have the opportunity of becoming acquainted with the more sedentary forms of Copepoda as we now are with the free-swimmers of this district.

The two Tables show the distribution of the Valencia and the "Argo" Copepoda arranged according to localities and dates.

REPORT on the Investigations carried on in 1895 in connection with the LANCASHIRE SEA-FISHERIES LABORATORY at University College, Liverpool.

By Professor W. A. HERDMAN, D.Sc., F.R.S., and
Mr. ANDREW SCOTT, Fisheries Assistant.

With Plates I.—V.

INTRODUCTORY.

THE work that can be done in the application of Zoological Science to the local Fishing industries seems spreading and increasing in amount each year; and the work of the past year, as may be seen from this Report, has opened up much fresh ground, and has been carried on not only in Liverpool and at Sea but also partly at Port Erin and to a slighter extent in the neighbourhood of Piel Island, near Barrow; and so has extended over the various parts of our northern district of the Irish Sea. In fact we may be regarded now as having, planned out, if not yet completely established, a system of Fisheries investigations which, although still on a small scale, will be able to cover the ground effectively and to cope adequately with the subject.

The central laboratory at University College, Liverpool, the Marine Biological Station at Port Erin, the steamer at sea, and the new branch laboratory now being fitted up at Piel Island can subdivide the work between them, and so render possible a wider range of observations. The finer microscopic and laboratory work, the comparison of results and the drafting of reports can only be carried on at a place like the Liverpool laboratory where microscopes, microtomes and other laboratory apparatus are available, where there are biological libraries to consult, and where there are other scientific workers to lend their help. The Biological Station at Port Erin affords facilities for practical work on the shore and for observations and experiments on the reproduction and rearing of young marine animals in tanks. Such observations will prepare the way for the proposed Sea-Fish Hatchery for which Port Erin seems pre-eminently fitted. The trawling observations, the examination of the spawning and feeding grounds, and the collection of statistics can only be carried out by the steamer at sea, under the direction of Mr. Dawson, as has been done in the past. Finally, the little laboratory now being fitted up at Piel Island will enable us to examine more systematically the great shell-fish beds of the northern district and to deal with fresh material brought in from that neighbourhood before it is preserved and sent on to the central laboratory at Liverpool.

Section I. of the following report, dealing with the foods of fishes, found by an investigation of the stomach contents, is in continuation of the work of previous years, and has been drawn up by Mr. Scott.

Section II., on the investigation of the tidal and other currents, which might affect the distribution of floating fish eggs and fish food, by means of "drift bottles," is a further account of the observations commenced last year and already discussed in a preliminary manner in the Ninth Annual Report of the Port Erin Biological Station. The results are now given more fully, and certain practical conclusions are drawn from them. I am myself responsible for this section.

A new line of enquiry has been commenced this year by sending Mr. Scott to examine some of the shell-fish beds of the district periodically, and bring back material consisting of shell-fish, of various sizes, and samples of the sea-bottom and sea-water, from which when fully examined in the laboratory a report can be drawn up on the condition of the beds. Section III., dealing with this investigation and giving lists of the organisms found on the beds has been drawn up by Mr. Scott. This can only be regarded as a first instalment of our report on the shell-fish beds, and the work will be continued during the present year. It need scarcely be pointed out that the branch laboratory on Piel Island will be of material assistance to us in examining the beds of the northern part of the district. Mr. Scott in the course of his examination of the mud from these mussel beds and of deposits from other parts of our district—notably the neighbourhood of Port Erin—has come upon a number of minute animals, chiefly Copepoda, not hitherto recognised as living in our district and some of them new to Science. These are described by Mr. Scott in Section IV. and are figured in Plates I. to V.

Section V. contains a preliminary account of the investigations now being carried on, partly in the Liverpool laboratory and partly at Port Erin, by Professor Boyce and myself, upon the conditions under which Oysters live healthily, and upon the supposed connection between oysters and disease—especially typhoid fever. It may be noted that in addition to the enquiry into the subject of the great "Oyster and Typhoid" scare, we have made many observations upon the different kinds of oysters grown or laid down in our neighbourhood, and the effect upon them of different kinds of water.

During last summer, I gave a Course of Free Lectures under the auspices of the Sea-Fisheries Committee and in accordance with the regulations of the University

Extension Scheme. The course was on "Our Edible Sea-Fish and Sea-Fisheries," and the lectures were delivered in the Zoology Theatre of University College on Monday evenings, commencing May 6th. The course was well attended, and the audience expressed much interest in the subject, many of them staying on at the conclusion of each lecture to examine the microscopic and other specimens and to ask questions. These lectures were not intended for, and were not attended by, fishermen *alone*, but were open to the general public; and I am convinced that it is fully as important for the future of Fisheries investigation and improvement and of just legislation in regard to the fisheries, that the general public should have opportunities of learning the truth in regard to the habits and life-histories of food fishes, and the inter-relations of animals in the sea, as it is that the fisherman himself should be instructed in such matters.

In addition to such public lectures, there is another method by which an educated public opinion upon Fishery questions can be formed, and that is by the establishment in each district of a technical museum or collection illustrating the local fisheries, the spawn and other stages in the life-history of the various fishes, their foods, their parasites, their diseases, and so on. Such a collection could now readily be formed, with very slight additional expenditure, at University College, in connection with the Fisheries Laboratory, and would then be available for consultation by fishermen, fishmongers, and all others concerned. It would be of constant use to ourselves, for comparison, in our fishery work; it ought to be of value to Fisheries inspectors and superintendents and to members of Sea-Fisheries Committees both in this neighbourhood and from other parts of the country; and it is the practical evidence that Fisheries experts from abroad especially desire to see when they come for information in regard to our local Fisheries and the conditions under which they are carried on.

Finally, such a technical museum of the Fisheries would naturally be made the scene and the means of object lessons and set demonstrations to the fishermen of the neighbourhood, and would probably be the most effective method of supplying technical instruction to that class of the community.

W. A. HERDMAN.

JANUARY, 1896.

SECTION I.

EXAMINATION OF FOOD IN FISHES' STOMACHS.

(By Mr. ANDREW SCOTT.)

WE have continued the examination of the stomachs of the various marine animals whose life-histories we are more intimately concerned with, chiefly from a fisheries point of view, and to which we have been paying considerable attention during the past few years, but as we now know fairly well what forms the chief food supply of these particular animals in our district, we do not deal with this part of the work in such an exhaustive manner as formerly and content ourselves by merely giving a summary of the results, noting any points of special interest connected with them.

During the past twelve months, from the beginning of January to the end of December, 1,540 stomachs of various marine animals from different parts of the district have been examined.

The following are the sources from which the stomachs have been obtained :—

Food fishes up to three inches	487
,, ,, above ,, ,, ...	493
Other fishes	20
Cockles	210
Mussels	230
Shrimps	100
	1,540

THE FOOD OF YOUNG FISHES.

The following summaries give the result of the examination of 487 stomachs of young food fishes, the differences between the year 1895 and the previous years, if any, being stated :—

Plaice (*Pleuronectes platessa*).

167 stomachs of young Plaice were examined, of these 36 were empty and 1 contained indistinguishable animal matter, leaving 130 to be accounted for as having recognisable matter.

Crustacea were found in 65 stomachs, exactly 50 %, and consisted of the remains of Amphipods and Copepods, 51 stomachs contained numbers of the Copepod *Jonesiella hyæna*, a species described by Mr. I. C. Thompson some years ago and which has since been found to enter very largely into the food supply of the young flat fishes.

Annelida were also found in 65 stomachs, 50 %. The stomachs examined last year and referred to in the Third Annual Report, gave a somewhat different result. Crustacea took first place with fully 63 %, Annelida second with 30 %, while 4 % of the stomachs contained Mollusca. So that it is quite clear that Crustacea and Annelida are the chief food supplying agents of the young plaice, and of the two, Crustacea is probably the more important.

Dab (*Pleuronectes limanda*).

272 stomachs of young Dabs were examined, of which 226 were found to contain no food and 11 contained food matter which was not recognisable, leaving only 35 to be accounted for.

Annelida were found in 34 stomachs, or fully 97 %. Echinoderms were found in 1 stomach, representing scarcely 3 %.

In last year's report 65 % of the stomachs of young Dabs examined contained Annelida, and 24 % Crustacea, so that Annelida appears to be by far the most important food supplying agent of the young Dabs, Crustacea occupying second place.

Flounder (*Pleuronectes flesus*).

5 stomachs of young Flounders were examined, of which 3 were empty and 2 contained the remains of Annelida.

Sole (*Solea vulgaris*).

5 young Soles were examined; and the stomachs were found to be empty.

Cod (*Gadus morrhua*).

10 stomachs of young Cod were examined, 7 of which were empty and 3 contained the remains of Crustacea.

Whiting (*Gadus merlangus*).

6 stomachs of young Whiting were examined, all of which were empty.

Sprats (*Clupea spratta*).

21 stomachs of young Sprats were examined, all of which were empty. Last year's report shows that out of 20 Sprats only 2 contained recognisable food, so that we have still to find out what the Sprats feed upon.

FOOD OF LARGER FISHES.

Plaice (*Pleuronectes platessa*).

153 stomachs of Plaice were examined, of which 43 were empty and 2 contained indistinguishable animal matter, leaving 108 to be accounted for.

Annelida were found in 68 stomachs, or nearly 63 %.

Mollusca were found in 36 stomachs, fully 33 %, and consisted of *Solen*, *Philine*, *Mactra*, *Cardium* and *Mytilus*. The shell-fish beds are without doubt a good feeding ground for the smaller flat and round fishes, as we frequently find the remains of Cockles and Mussels as well as Annelida in the stomachs of the fish caught in the vicinity of the shell-fish beds, so that the protection and cultivation of the more important shell-fish would be a means of increasing the food supply for the smaller sizes of the valuable food fishes.

Fish were found in 7 stomachs, or about $6\frac{1}{2}$ %, and consisted chiefly of sand eels.

Last year's report showed that Mollusca were the

most important food supplying agent of the Plaice, as fully 72 % of the stomachs contained the remains of various shell-fish, Annelida were second 22 % and Crustacea third with 8 %.

Dabs (*Pleuronectes limanda*).

176 stomachs of Dabs were examined, of which 70 were empty and 13 contained unrecognisable animal matter, leaving 93 to be accounted for.

32 stomachs contained Mollusca, or fully 34 %, the Mollusca consisted of the remains of *Buccinum*, *Cardium*, *Mactra*, *Mytilus*, *Philine* and *Nucula*. Here again we also find the smaller sizes of dabs, such as are caught in the shrimp nets, &c., feeding on the young Cockles and Mussels which they pick up on the shell-fish beds.

24 stomachs contained remains of Annelida, nearly 26 %.

23 stomachs contained remains of Crustacea, or nearly 25 %, and consisted of *Crangon*, *Pagurus*, *Portunus* and various Amphipoda.

6 stomachs contained remains of Echinoderms, nearly 7 %, and consisted chiefly of the sand starfish, *Ophioglypha*.

9 stomachs contained remains of fish, nearly 10 %, and were mostly composed of sand eels, but 1 stomach contained a number of fish eggs.

1 stomach contained remains of a Zoophyte.

Last year's report gave Annelida as first with 50 %, then Echinoderms, Mollusca, Crustacea and Zoophytes with 21 %, 20 % and 15 % respectively. In the previous year (1893) Mollusca were found to be the chief food.

Flounders (*Pleuronectes flesus*).

20 stomachs of Flounders were examined 15 of which were empty, the remainder, 5, contained fragments of Annelida.

Soles (*Solea vulgaris*).

40 stomachs of Soles were examined of which 27 were

empty, and 6 contained matter unrecognisable, so that only 7 out of the 40 contained food.

4 stomachs contained remains of Annelida.

2 stomachs contained remains of Crustacea.

1 stomach contained Mollusca.

Last year's report gave Annelida as being the most useful food supplying agent of this valuable food fish, Crustacea being second, so that it seems clear, that on the whole Annelida form a very important item in the food of the Sole.

It may be stated here that by far the greater number of Soles examined in this district are found to have no food in the stomachs.

Cod (Gadus morrhua).

30 stomachs of Cod were examined, of which 13 were empty, the remainder contained recognisable food matter.

Crustacea were found in 12 stomachs, or fully 70 %.

Fish were found in 4 stomachs, nearly 23 %.

Annelida were found in only 1 stomach.

Last year 90 % of the stomachs examined contained Crustacea, Fish being again second with 12 %, and Annelida third.

Whiting (Gadus merlangus).

74 stomachs of Whiting were examined, of which 38 were empty, and 5 contained unrecognisable food matter, leaving 31 to be accounted for.

Crustacea were found in 13 stomachs, or nearly 43 %.

Annelids were found in 3 stomachs, or fully 9 %.

Fish were also found in 3 stomachs, or fully 9 %.

Mollusca were found in 2 stomachs, or nearly 7 %.

In last year's report, Crustacea were found to occupy the first place with fully 73 %, Fish second with 24 % and Annelida third with 7 %.

Haddock (*Gadus æglefinus*).

32 stomachs of Haddock were examined of which 9 were empty, and 1 contained unrecognisable food matter, leaving 22 to be accounted for.

Crustacea were found in 12 stomachs, or fully 54 %.

Fish were found in 6 stomachs, or fully 27 %.

Mollusca were found in 3 stomachs, or fully 13 %.

Annelida were found in 2 stomachs, or fully 9 %.

From last year's Report it will be seen that Mollusca occupied the first place with fully 54 %, Echinoderms second with 21 %, Annelida and Crustacea third with fully 18 % each.

Thornback Skate (*Raia clavata*).

17 stomachs were examined of which 4 were empty, and 1 contained unrecognisable animal matter, leaving 12 to be accounted for as having contained recognisable food material.

Crustacea were found in 8 stomachs, or fully 66 %.

Mollusca were found in 3 stomachs, or 25 %.

Fish were also found in 3 stomachs, 25 %.

Last year's report shows that fully 97 % of the stomachs of the Skate contained Crustacea, Mollusca occupying second place with 16 %, and Fish third with 10 %, so that as far as our results go they show that the Thornback Skate in our district feed largely upon Crustacea such as *Crangon*, *Carcinus*, *Galathea*, *Hyas*, *Nephrops*, *Portunus*, *Pagurus* and a species of Amphipoda, probably *Ampelisca spinipes*, Boeck.

We have also examined the stomachs of a number of other more or less important food fishes, such as the " Lemon Sole " (*Pleuronectes microcephalus*), " Long Rough Dab " (*Hippoglossoides limandoides*), " Grey Gurnard " (*Trigla gurnardus*), " Starry Ray " (*Raia radiata*), " Grey Skate " (*Raia batis*), but not in sufficient numbers to make them worth while recording.

A number of the inedible fishes (fishes of no marketable value) such as the " Solenette " (*Solea lutea*), " Megrim " (*Arnoglossus laterna*), " Pogge " (*Agonus cataphractus*) etc., have been examined with the view of obtaining fresh information regarding their habits and food, so that we may have some idea as to what extent they compete with the more valuable food fishes of this district.

In the stomach of one of the Solenettes a young Sole, measuring $\frac{5}{8}$ of an inch, was found, which is the first direct evidence we have from this district of Solenettes feeding upon young Soles. Whether or not this happens to any great extent it is difficult to say.

CONCLUSION.

If we take into consideration the results of the four years work of examining stomachs and compare one year with another we find, as a rule, that each particular species of fish is fairly consistent in preferring one kind of animal as food.

There are times, however, when unusual animals may form a large proportion of the food, and this may well be due to a temporary scarcity of the usual foods or a temporary abundance of the forms substituted. Such variations in food matters may have considerable influence upon the movements of fishes within our district.

SECTION II.

THE DRIFT BOTTLES AND SURFACE CURRENTS.

(By Professor HERDMAN.)

IN last year's report the scheme for the distribution of drift bottles over the Irish Sea, for the purpose of helping to determine the set of the chief currents, tidal or otherwise, which might influence the movements of fish food and fish embryos, was fully explained. Since September, 1894, this work has been going on actively, and at the end of about twelve months over one thousand drift bottles in all had been set free. Many of them have been let out at intervals of ten minutes, or quarter of an hour, or twenty minutes (corresponding to distances of from 3 to 6 miles apart) from the Isle of Man boats when crossing between Liverpool and Douglas—a very convenient line of 75 miles across the middle of the widest part of our area, traversing the "head of the tide" or meeting place of the tidal currents entering by St. George's Channel and the North Channel. Others have been let off from Mr. Alfred Holt's steamers, in going round from Liverpool to Holyhead and in coming down from Greenock. Mr. Dawson on the Fisheries steamer "John Fell" has distributed a number along the coast in various parts of the district, and the Fisheries bailiffs have let off some dozens from their small boats. Other series have been set free at stated intervals during the rise and fall of the tide from the Morecambe Bay Light Vessel in the northern part of our area, north of the "head of the tide;" and, through the kindness of Lieutenant M. Sweny, R.N., a similar periodic distribution has taken place from the Liverpool North-West Light Vessel, to the south of the "head of the tide." Others, finally, have been despatched by Mr. Robert Harley, by Mr. R. L. Ascroft, by Mr. Andrew Scott and a few friends in other parts of the area from small boats and on our dredging expeditions, in some cases between the Isle of Man and Ireland. Altogether we have pretty well covered this northern area of the Irish Sea in our distribution of floating bottles.

Mr. Ascroft has also let off fifty larger and heavier bottles, champagne quarts weighted with sand so as to float almost entirely submerged, and with a post card attached to the end of the cork. Nearly 30 % of these have been returned; most were set free in the northern part of the district, and about 10 % have come south— *e.g.*, No. 34, set free off Duddon outer buoy (Cumberland) on 9th May, was found at Wallasey embankment (Cheshire) on the 18th of May—thus differing from our smaller bottles (see below) which have largely gone north. Possibly this difference in result may be due to the weighted champagne bottles having floated lower in the water or having been carried along near the bottom. Some of them are said to have sunk out of sight when set free, and one was trawled up from 12 miles S.E. of the Bahama Light Ship from a depth of 14 fathoms.

The first small bottles used, and the printed paper they contained, were described last year. We afterwards adopted a rather larger size of bottle, 8·5 cm. in length; and, after various postal difficulties and experiments, we hit upon a convenient size and thickness of private post card, which, ready stamped and addressed, and marked with a distinguishing letter or number, is rolled up in its bottle, and has printed on its back the following :—

For scientific enquiry into the currents of the Sea.

Whoever finds this is earnestly requested to write distinctly the DATE and LOCALITY, with full particulars, in the space below, and to put the card in the nearest post office.

[No. here]

LOCALITY, where found...................................

....

....

DATE, when found..

Name and address of sender............................

....

[No. here]

The number is marked with blue and black pencils in duplicate on opposite corners of the card, in case of one edge of the card getting worn by moisture; and the card is so rolled in the bottle that one of these numbers can be read through the glass, in order that a record may be kept of when and where each bottle is set free. Mr. Andrew Scott has collated these records with the particulars of finding of those bottles which have been recovered, and I am indebted to him for the details upon which the following statement of results is based.

Altogether out of the 1045 bottles distributed, over 440 or 42 per cent., more than two in five, have been subsequently picked up on the shore, and the paper, or post card, has been duly filled up and returned. We beg to thank the various finders of the bottles for their kindness in filling and posting the cards. They come from various parts of the coast of the Irish Sea—Scotland, England, Wales, Isle of Man, and Ireland. Some of the bottles have gone quite a short distance, having evidently been taken straight ashore by the rising tide; while others have been blown ashore by the wind, *e.g.*, two (post cards 211 and 214) let off near New Brighton stage on 9th October, 1895, the tide ebbing and the wind N.N.W., were found next day near the Red Noses, 1 mile to the west. Others have been carried an unexpected length, *e.g.*, one (No. 35), set free near the Crosby Light Vessel, off Liverpool, at 12.30 p.m., on October 1st, was picked up at Saltcoats, in Ayrshire, on November 7th, having travelled a distance of at least 180 miles* in thirty-seven days; another (H. 20) was set free near the Skerries, Anglesey, on October 6th, and was picked up one mile north of Ardrossan, on November 7th, having travelled

---

* More probably, very much further, as during that time it would certainly be carried backwards and forwards by the tide.

150 miles in thirty-one days; and bottle No. 1, set free at the Liverpool Bar on September 30th, was picked up at Shiskin, Arran, about 165 miles off, on November 12th. On the other hand, a bottle (J. F. 34) set free on November 7th, in the Ribble Estuary, was picked up on November 12th at St. Anne's, having only gone 4 miles.

We have not considered it necessary to give the particulars of every bottle that has been set free and afterwards recovered, but we have divided up our district into eight convenient areas in each of which a sufficiently large number of bottles has been set free, and the following table shows our results for these areas :—

AREA.	No. SET FREE.	No. RECOVERED.	APPARENT DIRECTION OF DRIFT.
West of I. of Man	84	33	West and North, mostly to Ireland.
East of I. of Man	115	29	Northward, mostly to Wigtonshire.
Central area	104	26	North-east, Cumbd., & N. Lanc.
North Wales	71	28	Mostly North-east.
Mersey area	173	95	24 N., 16 W., rest washed ashore.
N.W. Lt. Vessel	96	41	26 to North, 15 to West.
Ribble area	137	72	Half N. & N.E., rest ashore.
Morecambe B. & North	107	40	Mostly N.E. and E. (ashore).

It may be doubted whether our numbers are sufficiently large to enable us to draw very definite conclusions. It is only by the evidence of large numbers that the vitiating effect of exceptional circumstances, such as an unusual gale, can be eliminated. Prevailing winds, on the other hand, such as would usually affect the drift of surface organisms, are amongst the normally acting causes which we are trying to ascertain. Mr. W. E. Plummer of the Bidston Observatory has kindly given us access to his records of weather for the last twelve months, and we have noted opposite the bottles, from whose travels we are drawing any conclusions, an approximate estimate of the wind influences during the period when the bottle may have been at sea. There have been a few rather extraordinary journeys, *e.g.*, one let off in the middle of Port Erin Bay on April 23rd was found at Fleetwood on July 6th; another let off at Bradda Head on June 3rd was found on Pilling Sands (near Fleetwood) on July 24th.

It is important to notice that the bottles may support one another's evidence, those set free about the same spot often being found in the same locality, *e.g.*, out of a batch of 6 set free off New Brighton, on Oct. 9th, 1895, 5 have come back and all were found at about the same place.

Dr. Fulton, who has been conducting a similar inquiry by means of drift bottles, in the North Sea, for the Scottish Fishery Board, wrote to me some months ago that he was then having large numbers of his bottles returned to him from the Continent, chiefly Schleswig and Jutland. And he draws the conclusion, "There is no doubt that the current goes across, down as far as Norfolk—none of the bottles have been found south of Lincoln and none in Holland—and this will explain the presence of banks and shallows in the south and east, and the immense nurseries of immature fish there." Since then a detailed account* of these experiments on the Scottish coast has appeared, and it is interesting to compare our results with them.

Their experiments commenced on Sept. 21st, and ours on Sept. 30th, 1894. They report upon 729 bottles of which 159, or nearly 22 % have been returned, while we have distributed 1045 of which over 42 % have been found and recorded. The general result of the Scottish investigations is to show that most of the bottles are carried southwards in the North Sea along the east coasts of

* Thirteenth Ann. Report of Fishery Board for Scotland, Part III., p. 153.

Scotland and England as far as the Wash, and then east-ward to the Continent; so that the fish supply of a given area of the territorial waters may be derived not from the offshore spawning areas opposite, but from those situated further north—for example : the inshore waters of the Firth of Forth and St. Andrews Bay derive their main supplies not from the waters lying contiguous to them to the eastward, but from areas further north, such as the spawning grounds in the neighbourhood of the Bell Rock and those off the Forfarshire coast. In the case of our district, the drift is chiefly to the north and north-east (see below).

What is already well-known* in regard to the tidal streams or currents in the Irish Sea is that for nearly six hours after low-water at, say, Liverpool, two tidal streams pour into the Irish Sea, the one from the north of Ireland, through the North Channel, and the other from the south-ward, through St. George's Channel. Parts of the two streams meet and neutralise each other to the west of the Isle of Man, causing the large elliptical area, about 20 miles in diameter and reaching from off Port Erin to Carlingford, where no tidal streams exist, the level of the water merely rising and falling with the tide. The remaining portions of the two tidal streams pass to the east of the Isle of Man and eventually meet along a line extending from Maughold Head into Morecambe Bay. This line is the "head of the tide." During the ebb the above currents are practically reversed, but in running out the southern current is found to bear more over towards the Irish coast.

* All the accounts I have had access to seem based upon Admiral Beechey's observations published in the Philosophical Trans. for 1848 and 1851. Admiral Wharton, F.R.S,, the present Hydrographer to the Navy, has kindly informed me that Admiral Beechey took his observations by the direct method of anchoring his ship in various places and then observing the direction and force of the tide.

There is some reason to believe that, as a result of the general drift of the surface waters of the Atlantic and the shape and direction of the openings to the Irish Sea, more water passes out by the North Channel than enters that way, and more water enters by the South (St. George's) Channel than passes back, and that consequently there is, irrespective of the tides, a slow current passing from south to north through our district. The fact that so many of our drift bottles have crossed the " head of the tide " from S. to N., and that of those which have gone out of our district nearly all have gone north to the Clyde sea-area supports this view, which I learn from Admiral Wharton is *a priori* probable, and which is believed in by Mr. Ascroft and other nautical men in the district from their experience of the drift of wreckage.

It may be objected to our observations by means of drift bottles that they are largely influenced by the wind and waves, and are not carried entirely by tidal streams. Well, that is an advantage rather than any objection to the method. For our object is to determine not the tidal currents alone but the resulting effect upon small surface organisms such as floating fish eggs, embryos and fish food, of *all* the factors which can influence their movements, including prevalent winds. The only factors which can vitiate our conclusions are unusual gales or any other quite exceptional occurrences, and the only way to eliminate such influences is (1) to allow for them so far as they are known from the weather reports, and (2) to employ a large number of drift bottles and continue the observations over a considerable time. We have carefully considered the bearing of the weather records, and we think that the large number of bottles we have made use of, during the year, ought to enable us to come to some definite results. Our conclusions so far (January, 1896) then are:—

(1) A large number (over 42 %) have been stranded and found and returned,

(2) of these returned, only a small proportion (13 %) have been carried out of our part of the Irish Sea,

(3) nearly 12 % have crossed the "head of the tide," showing either the influence of wind in carrying floating bodies over from one tidal system into another, or the effect of that slow drift of water to the north referred to above,

(4) most of the bottles set free to the west of the Isle of Man have been carried across to Ireland, only a small number ($3\cdot8$ %) of them have got round to the eastern side of the Island and been carried ashore on the English coasts,

(5) the majority of the bottles set free off Dalby have gone to the Co. Down coast,

(6) a considerable number of bottles have been set free over the deep water to the east of the Isle of Man, where our more valuable flat fish spawn, and of those that have been returned the majority had been carried to the Lancashire, Cheshire and Cumberland coasts.

So we may reasonably conclude that *the embryos of fish spawning off Dalby would tend to be carried across to the Irish Coast, while those of fish spawning in the deep water on this eastern side of the Isle of Man would go to supply the nurseries in the shallow Lancashire and Cheshire bays, and very few would be carried altogether out of the district.*

The bearing of this conclusion upon the site of a Sea-Fish Hatchery for our district is obvious. It would not do to set the newly hatched larvæ free anywhere near the Lancashire or Cheshire coasts. Besides the muddiness and varying specific gravity of the water to which they would there be exposed, they would run too much risk of being carried straight ashore or stranded on sandbanks by wind and tide. They must be set free as nearly as possible where they would be found under natural conditions, and that is somewhere in the open deep water where the parent fish spawn. It is most satisfactory and re-assuring, then, to find, by these "drift bottle" experiments, that small objects set free in the surface layers of water to the east and south-east of the Isle of Man (where we can obtain the purest and most constant water for the purposes of a hatchery) are gradually carried over to the eastward; so that young flat fish hatched there, or set free there from a hatchery, by the time they have passed through their larval stages and are ready to take to the bottom in shallow water in order to search for Copepoda and other food matters, would find themselves in the Lancashire and Cheshire bays and estuaries which constitute our fish "nurseries."

SECTION III.

MUSSELS AND MUSSEL BEDS.

(By Mr. ANDREW SCOTT.)

WE have continued the examination of samples of the various kinds of shell-fish caught for sale in the Lancashire Sea-Fisheries district, but have nothing new to record beyond the fact that the results as to feeding and spawning confirm what has already been published in the former Reports.

With a view of securing further information, however, regarding the food, habits, life-histories, etc., of the shell-fish, we have begun to make periodic examinations of the various beds of the district. In connection with this investigation a visit was made to the Mussel beds at Piel, Duddon and Morecambe, about the middle of September last, the condition of the shell-fish was noted, and samples of material were collected for microscopic inspection at the University College Laboratory, with the following results. A short description of the beds is given first with remarks upon the Mussels, then come lists of the various animals observed either on the beds themselves or in the material collected from them.

Roosebeck Scars.

The Roosebeck Mussel Scars are situated about one and a half miles S.E. from Piel and are practically continuous with the shore, at low-water spring tides only a very narrow and shallow channel separates the inner scar from the outer, at high-water the beds are covered to a depth of several feet. The outer scar, which is evidently the most suitable for the production of this valuable shell-fish, is fully half a mile long and from one hundred to one hundred and fifty yards wide, and lies parallel to the shore, the whole area of this scar at the time the examination was made was covered with fine mud, reaching in some places to a depth of nearly two feet, but this mud, from the information supplied by Mr. Wright the head fishery officer of the Northern District, appears to have only settled down after the Mussels attached themselves to the hard ground, for until the Mussels appeared on the bed the ground was quite hard and free from mud. This experience is the same elsewhere, that Mussels tend to accumulate mud and so raise the level of the bed.

The outer scar is clearly a much better rearing place for Mussels than the inner one as the present condition of the shell-fish show, for while the Mussels on the outer scar which only settled down in April, 1895, had reached in the majority of cases to nearly three-quarters of an inch in about a period of five months, those on the inner scar deposited at the same time had scarcely attained to one-quarter of an inch in length when the examination was made. The Mussels on the outer scar are very numerous being so closely packed together in some places that there appears to be scarcely any space left for further growth of the shell-fish, but as they have now no firm attachment owing to their having to keep moving as the mud accumulates, they are liable to be washed off by any heavy sea from the southward, and carried off into the deeper water outside the scar from whence they can only be obtained by raking. The shells are quite clean and free from barnacles.

It is difficult to explain why the Mussels on the outer sear should grow so much faster than those on the inner scar as the conditions of the bottom must have been much alike in structure and wealth of food supply before the spat settled down. Possibly the rapid growth of the Mussels on this scar may be due to the large quantities of fresh water that pass over it when the tide is out, little or none of which appears to reach the Mussels on the inner scar. This water will bring down diatomaceous and other material collected on its passage across the shore, which will be deposited when it comes into contact with the sea and will accumulate in considerable quantities just at low-water mark, forming a splendid rearing ground for the microfauna that forms the principle source of food of the Mussel. Whether this be the true explanation or not, the rapid growth of the Mussels on the outer scar is without doubt due to some important cause which does not affect the inner scar, and which can only be found out by further investigation.

FAUNA OF THE SCAR.

Besides Mussels, the usual animals that live in association with them are found here, the chief enemies being the boring Mollusca, which perforate the Mussel shells and extract the juices and flesh of the animal, such as the "Dogwhelk," *Purpura lapillus*, L. The "Whelk," *Buccinum undatum*, L., the common "Edible periwinkle," *Littorina litorea*, L., and a few common shore crabs, *Carcinus moenas*, were also present, but none of them in any great quantity. The examination of the mud yielded seventeen species of Foraminifera representing ten genera, ten species of Ostracoda representing six genera, and eleven species of Copepoda representing ten genera. The mud also contained fragments of shells, spines of *Spatangus*, sponge spicules, a few diatoms and a considerable quantity of vegetable debris. The following list gives the names of the species of Ostracoda, Foraminifera and Copepoda. The Ostracoda and Foraminifera from the Mussel beds at Piel, Duddon and Morecambe have been identified by Mr. Thomas Scott, F.L.S., and Dr. G. W. Chaster.

FORAMINIFERA.

Lagena clavata, d'Orb.
,,   *lœvis*, Mont.
,,   *striata*, d'Orb.
,,   *williamsoni*, Alcock.
,,   *squamosa*, Mont.
,,   *sulcata*, W. and J.
Polystomella striato-punctata, F. and M.
Polymorphina lactea, W. and J.
Planorbulina mediterranensis, d'Orb.
Nonionina depressula, W. and J.
Rotalia beccarii, L.

Cornuspira involvens, Reuss.
Biloculina depressa, d'Orb.
Bulimina aculeata, d'Orb.
Miliolina oblonga, Mont.
    ,,   *seminulum*, d'Orb (juv.).
    ,,   *subrotunda*, Mont.

OSTRACODA.

Cythere pellucida, Baird.
Cytherura angulata, Brady.
    ,,   *nigrescens* (Baird).
    ,,   *sella*, G. O. Sars.
Cytherideis subulata, Brady.
Loxoconcha impressa, Baird.
    ,,   *tamarindus* (Jones).
Sclerochilus contortus, Norman.
Paradoxostoma abbreviatum, G. O. Sars.
            flexuosum, Brady.

COPEPODA.

Temora longicornis (Müller).
Paracalanus parvus (Claus).
Canuella perplexa, T. and A. Scott.
Ectinosoma curticorne, Boeck.
Euterpe acutifrons, Dana.
Ameira exigua, T. Scott.
Laophonte serrata (Claus).
    ,,   *lamellifera* (Claus).
Cletodes propinqua, Brady and Robertson.
Harpacticus chelifer (Müller).
Idya furcata (Baird).

SCARFHOLE SCAR, NEAR DUDDON.

This Mussel bed is fully four miles from Barrow-in-Furness and is situated nearly opposite the north end of

Walney Island. Like the Roosebeck Scars it is almost
continuous with the shore and at low-water of spring
tides is nearly dry, but is covered to a depth of several feet
at high-water. The bottom is hard and consists of sand,
stones and Mussel shells, quite different from that at
Roosebeck, consequently the Mussels have a much firmer
hold of the bottom and are not so easily washed off.

The Mussels on this bed are numerous, large and in
good condition, but are covered with barnacles.

FAUNA OF THE SCAR.

Besides the Mussels, *Buccinum, Purpura, Littorina*
and *Carcinus* were also found on the bed. An examination
of the mud yielded thirteen species of Foraminifera
representing eight genera, five species of Ostracoda
representing four genera, and eight species of Copepoda
representing seven genera. The mud also contained
fragments of shells, spines of *Spatangus*, sponge spicules
and a few diatoms. The following is a list of the Ostracoda, Foraminifera and Copepoda :—

FORAMINIFERA.

Lagena lævis, Mont.
    ,,   *striata*, d'Orb.
    ,,   *sulcata*, W. and J.
    ,,   *williamsoni*, Alcock.
Polystomella striato-punctata, F. and M.
Planorbulina mediterranensis, d'Orb.
Nonionina depressula, W. and J.
Bulimina elegans, d'Orb.
    ,,   *pupoides*, d'Orb.
Miliolina seminulum, L.
    ,,   *subrotunda*, Mont.
Nodosaria communis, d'Orb.
Haplophragmium canariense, d'Orb.

OSTRACODA.

Cythere confusa, Brady and Norman.
,,　　*lutea* (O. F. Müller).
Loxoconcha tamarindus (Jones).
Sclerochilus contortus, Norman.
Paradoxostoma variabile, Baird.

COPEPODA.

Canuella perplexa, T. and A. Scott.
Tachidius brevicornis, Müller.
Delavalia palustris, Brady.
Laophonte curticauda, Boeck.
　　,,　　*intermedia*, T. Scott.
Nannopus palustris, Brady.
Platychelipus littoralis, Brady.
Thalestris harpactoides, Claus.

MORECAMBE MUSSEL BEDS.

These beds by far the most valuable and extensive of the northern district, if not of the whole area under the control of the Lancashire Sea-Fisheries Committee, extend from a little way south-east of the town of Morecambe, to a considerable distance beyond the village of Heysham and are partly continuous with the shore. In front of Heysham, they are separated at low water into three separate banks, the outer one tailing off into the deep-water at the entrance to Heysham Lake. The Mussel scar separates the deep-water into two distinct channels, the outer being Morecambe Channel, the fairway to Morecambe, while the inner is known as Heysham Lake, which at low-water is open only at the south-west end. At high-water the beds are completely submerged and consequently can only be worked upon at low-water or at half tide by means of long mussel rakes. At the time the examination was made no "musselling" was going on.. Hedge balks were erected on the tops of the various banks in which a number of codling, plaice, and flounder were being taken.

The area of the bed is composed almost entirely of the deserted sand tubes of *Sabella* which afford a firm hold to the growing Mussels and no doubt also a good food supply from the diatoms and other microscopic animals which feed on the decaying matter about the tubes. Scattered over the beds are numerous large boulders all more or less covered with deserted sand tubes.

The Mussels here are healthy and in good condition showing evidence of an abundant food supply and of conditions favourable to a rapid growth of the shell-fish. They are quite free from barnacles.

Fauna of the Scar.

Along with the Mussels, there were also a few *Buccinum*, *Purpura*, *Littorina* and *Carcinus*. On microscopic examination the mud was found to contain nineteen species of Foraminifera representing eleven genera, eight species of Ostracoda representing five genera, and nine species of Copepoda (one of which appears to be new) representing eight genera. The mud also contained a few fragments of shells, spines of *Spatangus*, diatoms and a quantity of vegetable debris. The following is a list of the Ostracoda, Foraminifera and Copepoda :—

Foraminifera.

Lagena striata, d'Orb.
    ,,   *williamsoni*, Alcock.
    ,,   *clavata*, d'Orb.
    ,,   *semistriata*, Will.
    ,,   *sulcata*, W. and J.
    ,,   *hexagona*, Will.

„ *globosa*, Mont.
Polystomella striato-punctata, F. and M.
Planorbulina mediterranensis, d' Orb.
Nonionina depressula, W. and J.
„ *stelligera*, d'Orb.
Rotalia beccarii, L.
Biloculina depressa, d'Orb.
Miliolina subrotunda, Mont.
„ *trigonula*, Lamarck.
Haplophragmium canariense, d'Orb.
Truncatulina lobatula, W. and J.
Bolivina plicata, d'Orb.
Virgulina schreibersiana, Cz.

OSTRACODA.

Cythere confusa, Brady and Norman.
„ *robertsoni*, Brady.
Cytherura cellulosa, Norman.
„ *striata*, G. O. Sars.
„ *sella*, G. O. Sars.
Cytheridea elongata, Brady.
Loxoconcha guttata, Norman.
Sclerochilus contortus, Norman.

COPEPODA.

Longipedia minor, T. and A. Scott.
Canuella perplexa, T. and A. Scott.
Ectinosoma curticorne, Boeck.
Bradya minor, T. and A. Scott.
Laophonte intermedia, T. Scott.
Cletodes propinqua, Brady and Robertson.
Platychelipus littoralis, Brady.
Idya furcata (Baird).
„ *elongata*, n.sp.

CONCLUSION.

The whole area of Morecambe Bay, with its numerous banks and the estuaries of the Wyre and Lune rivers and Morecambe, Grange, Ulverston, Barrow and Duddon Channels, by a little artificial cultivation might easily be turned into a vast rearing ground for Mussels, as everywhere the conditions seem most favourable for the production and rearing of this valuable shell-fish. All that is necessary is to remove some of the seed Mussels from the beds where they are too crowded to places where they are few in number or entirely absent. Many of the Mussels on the various beds under the present conditions are certain to be lost, through having too little room to grow, when consequently they either get choked off and perish miserably or become so stunted in their growth as to be of little or no marketable value either as bait or food.

Under the present conditions the beds are allowed to seed themselves and consequently it takes much longer for the Mussel to spread over any considerable area, than if the young Mussels were transplanted from places on the present beds where they are too close, to new ground. By this means the Mussels would certainly reach a marketable size much sooner than if left to look after themselves, and a convenient time for moving the seedlings would be when the close season begins, as they would be less liable to be disturbed than if done at any other period, and by the time the close season had expired, they would no doubt have firmly established themselves under their new conditions.

We propose during the coming year to continue this investigation by doing some more work at Piel when the temporary laboratory is fitted up, in examining the Mussel beds of the middle and southern district. The southern beds can easily be examined from the central laboratory at University College, but the examination of the beds at Lytham and Fleetwood would be better done from Piel. An examination will also be made of the Cockle beds of the district as far as time permits.

The question has been raised whether Mussels which are no longer quite young and which have been torn off from their first supports, or which have been detached from larger Mussels, as in separating up bunches, are able to fix themselves anew. It is known to Zoologists that Mussels are able to produce byssus threads at any time and so re-attach themselves to any foreign object, consequently there can be no doubt that the smaller individuals torn off a bunch, will, if thrown back promptly into suitable ground, be able to spin fresh byssus round neighbouring objects and so become anchored. In order to settle the question quite definitely we have made some observations in the laboratory during the last year, and especially quite recently. One of our tanks had a number of young Mussels, varying in size from half an inch to an inch and a half in length, put in last June. These after climbing about the sides of the tank for some time attached themselves by means of byssus threads in various positions. They were occasionally detached, and were usually found re-attached after a few days. That has gone on during the last eight months. The Mussels that now survive in the tank have increased considerably in size one being over 2 inches and another 3 inches in length. The former of these was torn off from the side of the tank lately and was thrown into the middle. In two days it was found re-attached by byssus to the glass. Again, on February 4th the largest Mussel, measuring 3 inches in length, $1\frac{1}{2}$ inches in breadth, and $1\frac{1}{4}$ inches in thickness,

-and weighing 52½ grammes, was torn off from its attachment to the glass and placed on the sand in the bottom of the tank. In four days it had re-attached itself to the glass by means of byssus threads. This shows, if any further demonstration was really required, that even Mussels which have attained to large size have the power of spinning fresh byssus threads by which they become anchored to surrounding objects.

SECTION IV.

DESCRIPTION OF NEW AND RARE COPEPODA.

(By Mr. ANDREW SCOTT.)

Family HARPACTICIDÆ.

Sunaristes paguri, Hesse.

This rather peculiar and interesting species was obtained by washing the shells of *Buccinum* inhabited by the hermit crabs *Pagurus bernhardus*, collected in the trawl-net of the steamer while working at the mouth of the Mersey estuary on the 23rd of July, 1895. It seems to be a comparatively rare species and so far as is known this is only the third time it has been found in British waters. From our present knowledge of its distribution it appears to be confined to areas having large volumes of brackish water passing over the bottom, and has not been found in pure sea-water.

Sunaristes paguri is not unlike *Canuella perplexa* in general appearance but is readily distinguished from that species by the structure of the various appendages, especially the antennules and second pair of swimming feet of the male.

Stenhelia herdmani, n. sp. Pl. I., figs. 1—11.

Description of the species.—Female. Length 1·43 millim. ($\frac{1}{17}$th of an inch. Body moderately stout ; rostrum prominent and curved. Antennules long and slender, eight-jointed ; the first, second, fourth and eighth joints longer than the others, the fifth joint being the smallest of the series ; the second, third and fourth joints have each a tuft of setæ on their upper distal margins. The proportional lengths of the various joints are as follows :—

$$\frac{14 \cdot 14 \cdot 8 \cdot 10 \cdot 5 \cdot 6 \cdot 7 \cdot 11}{1 \quad 2 \quad 3 \quad 4 \quad 5 \quad 6 \quad 7 \quad 8}$$

Antennæ moderately stout, secondary branch small and slender, two jointed ; basal joint elongate narrow with one seta on its upper distal end, second joint short, about one third of the length of the first and furnished with two terminal setæ. Mandibles large and well developed, the broad biting part armed with a few large teeth and a number of smaller ones ; mandible palp comparatively large, consisting of a one-jointed basal part which carries at its lower extremity two branches, one large and one small, the smaller of the two being two-jointed, whilst the larger one is composed of a single joint. Masticatory portion of the maxillæ furnished with a number of strong teeth, palp two branched, the outer one bearing three setiferous lobes. Anterior foot-jaws furnished with one large terminal claw and three digitiform setose tubercles. Posterior foot-jaws stout, of moderate length and furnished with a strong, slightly curved terminal claw at the base of which are two setæ ; the basal joint of the foot-jaw has four small ciliated tubercles on its lower side, while the second joint has a row of fine cilia on its upper margin and a row of stronger cilia on its lateral surface a little way down from the upper margin, there are also two plumose setæ on the upper margin of the joint. First pair of swimming feet somewhat similar to those of *Stenhelia ima*, Brady; basal joints of the inner branches nearly as long as the entire outer branches, second joint about half the length of the third which is less than one third the length of the long basal joint. Outer branches of the second, third and fourth pairs elongate, inner branches much shorter, those of the fourth pair only reaching to the end of the second joint of the outer branches. Fifth pair of feet large and well developed, inner branches considerably larger than the outer ones, with a subtriangular apex bearing five plumose setæ, two on the outer angle close together and three arranged at regular intervals along the inner margins; outer branches subovate, bearing six setæ on the external distal margins, the second seta from the inside is considerably longer than any of the others. Caudal stylets about as long as broad and about half the length of the last abdominal segment.

Habitat, 1 mile off Spanish Head, Isle of Man, in neritic material dredged from a depth of 16 fathoms, October 27th, 1895.

Remarks.—This large and well marked species though somewhat like *Stenhelia ima* in general appearance is readily distinguished from it and the other known members of this genus, by the form and armature of the fifth pair of feet, and by the structure and proportional lengths of the antennules.

Stenhelia similis, n. sp. Pl. I., figs. 12—25.

Description of the species.—*Female.* Length 1 millim. ($\frac{1}{25}$ of an inch). Body elongate, moderately robust; rostrum prominent and curved with a bifid apex. Antennules long and slender, sparingly setiferous, the second joint longer than any of the others and slightly contracted near the middle, but expanding again towards the distal end, third, fifth, sixth and seventh joints small, the others of moderate length as shown by the formula :—

$$\frac{13 \,.\, 24 \,.\, 7 \,.\, 12 \,.\, 5 \,.\, 6 \,.\, 6 \,.\, 10}{1 \quad 2 \quad 3 \quad 4 \quad 5 \quad 6 \quad 7 \quad 8}$$

Antennæ well developed, secondary branch three-jointed, second joint very small, terminal joint fully half the length of the basal one and furnished with two setæ on its apex, one large and spiniform and one very small; one seta springs from near the middle of the upper margin of the terminal joint, the basal joint bears one seta on its upper distal angle. Mandibles furnished with several strong and serrated teeth on the biting parts, mandible palp consisting of a basal part carrying two branches, the inner branch which is smaller than the outer is two-jointed, both branches are furnished with a number of setæ on the apex and upper margins, the basal part has three terminal plumose setæ, and a curved row of short spines on its lateral surface. Maxillæ somewhat similar to those of *Stenhelia herdmani*. Posterior foot-jaws slender and furnished with a short curved claw, basal joints short and furnished with three small plumose setæ on the upper distal margin, second joint fully three times longer than broad and bearing a few cilia and one seta on its upper margin, there are also a few spines on its lateral surface. The first four pairs of swimming feet are nearly as in *Stenhelia ima*, the joints of the outer branches of the first pair are subequal, basal joint of the inner branches nearly as long as the entire outer branch, second joint small and about half the length of the third which is about half the length of the basal joint, the apex of the third joint is furnished with one short stout spine and two plumose setæ, one long and one short. Fifth pair of feet large, inner branches broad and triangular, bearing five short plumose setæ from the middle of the inner margin to the apex ; outer branches elongate ovate, about two-thirds the length of the inner, proximal half of the outer margin ciliated, inner margin slightly ciliated towards the distal end, apex and distal half of the outer margin furnished with six setæ, the second from the inner part of the apex considerably longer than the others. Caudal stylets rather shorter than broad and about one-third the length of the last abdominal segment.

Male. Antennules ten-jointed, fourth and sixth joints very small. Swimming feet, with the exception of the second pair, similar to those of the female. Inner branches of the second pair two-jointed, second joint bearing at the apex two strong and slightly curved spines, the inner spine which is slightly longer than the outer one, becomes distinctly bifid at the middle. The form of the fifth pair of feet is somewhat similar to those of the female, but smaller and furnished with fewer setæ, the inner branches have only two setæ which are placed on the apex, the outer branches have two setæ on the outer distal margin, the lower one being stout and spiniform, two setæ on the middle of the inner margin and one seta on the apex.

Habitat, 1 mile off Spanish Head, Isle of Man, in neritic material dredged from a depth of 16 fathoms. A considerable number of specimens were obtained.

Remarks.—This species comes near *Dactylopus tenui-remis,* but can easily be distinguished from it by the structure and proportional lengths of the antennules, the length and armature of the inner branches of the first feet, and also by the structure of the fifth feet.

Stenhelia reflexa, T. Scott.

[T. Scott, Thirteenth An. Rep. Fish. Board for Scot., pt. III., p. 166, 1895.]

A few specimens of this *Stenhelia* were obtained from dredged material collected off Port Erin in June, 1895.

Ameira gracile, n. sp.   Pl. II., figs. 1—11.

Description of the species.—Female.   Length ·5 millim. ($\frac{1}{50}$th of an inch).   Body elongate and slender, rostrum small and inconspicuous.   Antennules long and very slender, seven-jointed; second and fifth joints longer than any of the others, fourth joint very short, the second, third and fourth joints have each a tuft of long setæ on the upper distal margins, the following formula shows the proportional lengths of the joints :—

$$\frac{9 \cdot 18 \cdot 10 \cdot 4 \cdot 13 \cdot 8 \cdot 9}{1 \quad 2 \quad 3 \quad 4 \quad 5 \quad 6 \quad 7}$$

Antennæ slender, three-jointed, secondary branch small, two-jointed, the second joint very small.   Mandibles elongate narrow, apex obliquely truncate and armed with a number of teeth, mandible palp with a distinct basal part, narrow at the base but somewhat dilated towards the apex to which is attached a one-jointed elongate narrow branch.   Posterior foot-jaws moderately robust and armed with a strong terminal claw, lower margin of the second joint furnished with a row of fine cilia.   First pair of swimming feet elongate and slender, basal joint of the inner branches nearly as long as the entire outer branch, second joint about one-fourth the length of the basal joint and fully half the length of the third joint.   Outer branches of the second, third and fourth pairs elongate three-jointed, inner branches also three-jointed but shorter than the outer branches.   Fifth pair of feet foliaceous, the inner branch produced into a subtriangular lobe which reaches to about the middle of the outer branch and furnished at the apex with a stout setiform spine and a small seta, outer branch oblong ovate in shape, the greatest breadth being very nearly half the length, furnished with three setæ on the outer margin, one on the apex and one on the inner distal margin, both the inner and outer margins are clothed with fine cilia. Caudal stylets long and narrow, being about five times longer than broad and nearly twice the length of the last abdominal segment.

Male. Antennules ten-jointed, fifth and sixth joints very small, hinged between the third and fourth joints and also between the seventh and eighth joints. The form of the fifth pair of feet is somewhat similar to those of the female, but the inner branch is much smaller.

Habitat, 1 mile off Spanish Head, Isle of Man, in neritic material dredged from a depth of 16 fathoms, a number specimens were obtained.

Remarks.—This species in general appearance is not unlike *Ameira longicaudata* but is readily distinguished from it by the shape of the cephalothoracic segment and on dissection by the characters described above. Nearly all the specimens obtained had the last three joints of the antennules broken off.

Ameira reflexa, T. Scott.

[T. Scott, Twelfth An. Rep. Fish. Board for Scot., pt. III., p. 240, 1894.]

One or two specimens of this *Ameira* were obtained from the shelly deposit dredged 1 mile off Spanish Head, Isle of Man, depth 16 fathoms. The species is easily distinguished from the other members of this genus by the structure of the inner branches of the first pair of swimming feet and also by the fifth pair of feet.

Canthocamptus palustris, Brady. Pl. II., figs. 12—23.

[Brady, Monograph Brit. Copep., Vol. II., p. 53, 1880.]

A considerable number of specimens of a copepod apparently belonging to this species were washed from mud adhering to samples of Mussels (*Mytilus edulis*) sent from the St. Annes Mussel beds near Lytham, one of the samples was from that part of the bed which never becomes dry at low-water, and was obtained by means of a "mussel rake," it was from this sample that the first specimens were obtained, other samples sent later on in the year also contained numbers of specimens.

The specimens differ a little from the figures given by Dr. Brady in his "monograph," especially in the length of the basal joint of the first pair of swimming feet and also in the shape of the fifth pair of feet of the female.

Mesochra macintoshi, T. and A. Scott.

[T. & A. Scott, An. & Mag. Nat. Hist., Ser. 6, Vol. XV., p. 53, 1895.]

A number of specimens of this species were obtained from the shelly material dredged 1 mile off Spanish Head, Isle of Man, from a depth of 16 fathoms. The slender appearance of the species along with the structure and armature of its various appendages, enable it to be readily distinguished from the other members of the genus.

Tetragoniceps trispinosus, n. sp. Pl. II., figs. 24 and 25; III., figs. 1—6.

Description of the Species.—Female. Length ·5 millim. ($\frac{1}{50}$th of an inch). Body elongate cylindrical, tapering gently towards the posterior end, rostrum small and triangular in shape. Antennules long and slender, six-jointed and sparingly setiferous, the basal joint is considerably longer than any of the others, fifth joint very small, about half the length of the fourth; the proportional lengths of the joints are as shown by the following formula:—

$$\frac{28 \cdot 13 \cdot 14 \cdot 8 \cdot 4 \cdot 16}{1 \quad 2 \quad 3 \quad 4 \quad 5 \quad 6}$$

Antennæ of moderate length and three-jointed, secondary branch small and rudimentary, consisting of a single seta attached to the lower margin of the second joint of the primary branch at a distance of about one-third from the base. Posterior foot-jaws small, with a strong curved claw as long as the joint to which it is attached. Both branches of the first pair of swimming feet two-jointed, outer branches small, the joints subequal and reaching to about the middle of the basal joint of the inner branch; inner branches long and slender, basal joint nearly twice the length of the entire outer branch and fully seven times longer than broad, a moderately long seta springs from near the base of the inner margin. Second joint short and narrow, fully one-fourth the length of the basal joint, furnished at its apex with a short curved seta, a seta of considerable length springs from near the middle of the inner margin. Outer branches of the second; third and fourth pairs of feet elongate, three-jointed, inner branches short and narrow, one-jointed, in the fourth pair the inner branches are only about one-third the length of the basal joint of the outer branches and furnished at the apex with three short setæ. Fifth pair of feet small, one branched and divided into two distinct portions, an inner which is produced into an elongate curved spiniform apex devoid of setæ and an outer tubercle-like process which arises from near the base of the elongate portion furnished with two short stout setæ and one long slender hair. Caudal stylets elongate narrow, slightly divergent, tapering to an acute apex and about twice the length of the last abdominal segment; on the inner margin of each stylet at a distance of about one-third from the apex there arises a single seta which is fully two-thirds the length of the animal and having a slighly thickened base. Anal operculum semi-circular in shape and produced into three spines, a median and two lateral.

 Habitat, 1 mile off Spanish Head, Isle of Man, in neritic material, dredged from a depth of 16 fathoms. Only two specimens were observed.

 Remarks.—This species though placed in the genus

Tetragoniceps differs somewhat from the generic description given in the Monograph of the British Copepoda,
especially in the number of joints in the outer branches
of the first pair of feet and in the inner branches of the
second, third and fourth feet, but as the mouth organs
have not been satisfactorily worked out, it is perhaps
better meanwhile to place it under the genus *Tetragoniceps*
its nearest ally rather than institute a new genus for its
reception.

Tetragoniceps consimilis, T. Scott.

[T. Scott, Twelfth An. Rep. Fish. Board for Scot., pt.
III., p. 244, 1894.]

A few specimens of this species were obtained from the
material dredged 1 mile off Spanish Head, Isle of Man,
from a depth of 16 fathoms, it closely resembles *Tetragoniceps bradyi* in general appearance as well as in a few
structural details, but differs from it in the absence of the
strong hook on the second joint of the antennules, in the
inner branches of the first pair of feet being three-jointed
and in the fifth pair being composed of two distinct
branches.

Laophonte propinqua, T. and A. Scott.

[T. & A. Scott, An. & Mag. Nat. Hist., Ser. 6, Vol.
XV., p. 460, 1895.]

A few specimens of this species were obtained from
material washed from sponges collected by Dr. Hanitsch
at Port Erin, Isle of Man, in August, 1894; it is not
unlike *Laophonte denticornis* at first sight but on closer
examination is found to differ very markedly, not only
from that species, but from any of the other known
members of the genus.

Laophonte intermedia, T. Scott.

[T. Scott, Thirteenth An. Rep. Fish. Board for Scot.,
pt. III., p. 168, 1895.]

This species was obtained from the same material as the last, and also from the mussel beds at Duddon̄ and Morecambe, it appears to be intermediate between *Laophonte lamellata* and *Laophonte hispida* but is quite distinct from either of them, the sub-conical form of the stylets alone enable it to be easily recognised when mixed up in a collection of Copepoda along with *L. lamellata* and *L. hispida*.

Pseudolaophonte, n. gen.

Description of the genus.—*Pseudolaophonte* resembles *Laophonte*, Philippi, in the structure of the antennules and antennæ; the mandibles, maxillæ and foot-jaws, and the first pair of swimming feet, but differs from that genus in the structure of the second and third pairs; the second pair of swimming feet consist of a single one-jointed branch, and the outer and inner branches of the third pair are each composed of two joints. The fourth and fifth pairs of feet are somewhat similar to those of *Laophonte*.

Pseudolaophonte aculeata, n. sp. Pl. III., figs. 7—23.

Description of the species.—*Female.* Length 1 millim. ($\frac{1}{25}$th of an inch). Body seen from above elongate narrow, of nearly equal breadth throughout, all the segments are more or less angular in shape and furnished with a row of short teeth on their posterior margins; surface of all the segments clothed with minute cilia; rostrum small and inconspicuous, with a small hair on each side of the base. Antennules moderately stout, four-jointed, first and third joints longer than the other two, the fourth joint being the smallest, the basal joint has a row of blunt pointed teeth on its upper margin and three rows on its lateral aspect, the middle row being the longest; a stout tubercle with a quadri-dentate apex arises from near the middle of the lower margin; second joint furnished on its lower margin with a strong slightly curved tooth which reaches to near the middle of the basal joint, and forms with the dentate tubercle of that joint, a powerful grasping apparatus; the third joint is covered with minute spines for about three-fourths of its length, the remaining fourth being covered with fine cilia, the fourth joint is also covered with cilia and has the lower distal part produced into a strong spine, the following formula shows the proportional lengths of the joints :—

$$\frac{17 \cdot 11 \cdot 16 \cdot 6}{1 \quad 2 \quad 3 \quad 4}$$

Antennæ two-jointed and of moderate size, with a small one-jointed secondary branch arising from near the middle of the lower margin of the basal joint and furnished with four setæ. Mandibles small, with a few serrated teeth on the truncate apex, mandible palp very small, with ciliated margins and bearing three setæ on the apex. Maxillæ and foot-jaws somewhat similar to those of a typical *Laophonte*, the second joint of the posterior foot-jaw long and slender, being about four times longer than broad, the terminal claw is also long and slender and is considerably longer than the second-joint. First pair of swimming feet similar to those of a typical *Laophonte*, outer branch composed of two joints. Second pair of swimming feet rudimentary, consisting of a single one-jointed branch, bearing three setæ at the apex, the innermost being longer than the other two. Both branches of the third pair of feet two-jointed, the inner branch being slightly shorter than the outer. The fourth pair of feet has the outer branch three-jointed and the inner two, the basal joint of the outer branch is nearly as long as broad and is equal to the combined lengths of the second and third joints, the first and second joints have each one stout ciliated spine on the outer distal angle, the second joint which is very narrow, is produced on the inner margin into a hook-like process furnished with a short seta, the third joint has three strong spines on the outer margin and apex, inner branches short, reaching to about the middle of the outer branch, the second joint is furnished with three short setæ on its apex. Fifth pair of feet large and foliaceous, inner branch triangular in shape, ciliated on the inner margin and covered with a number of more or less curved rows of cilia, the branch is also furnished with five moderately stout plumose setæ on its inner margin and apex; outer branch broadly ovate, and fully half the size of the inner branch, it is also covered with rows of cilia and bears five short stout plumose setæ on its apex. Caudal stylets elongate narrow, of moderate length, about three times longer than broad and slightly longer than the last abdominal segment; bearing on the inner angles of the apex, a short stout curved spine and near the middle the dorsal surface, a slightly shorter spine and a seta, the outer margins are furnished with two short setæ, the apex also bears two setæ, one of which is very long. Anal operculum produced into a short stout spine.

Male. Antennules six-jointed, first and second joints like those of the female, third and sixth joints very small, fourth joint considerably dilated. Mouth organs similar to those of the female. The first and second feet are also similar to to those of the female. The basal joint of the outer branches of the third pair of feet has a strong curved spine on its outer distal angle which is nearly twice the length of the joint itself and extends considerably beyond the end of the second joint, second joint of the inner branches produced into a curved spine which reaches to beyond the end of the outer branch, both branches of the third pair two-jointed. The fourth pair of feet has the outer branch three-jointed and the inner two; the basal joint of the outer branches is longer than the combined lengths of the second and third joints and bears a strong spine on its outer distal angle, second and third joints of the outer branch of about equal length; inner branches very short reaching to about the middle of the basal joint of the outer branch, basal joint of the inner branch very small and only about one-fourth the length of the second joint. Fifth pair small, inner branch not produced, furnished with two plumose setæ on its apex, the inner one being three times longer than the outer; outer branch elongate narrow, bearing at its apex three stout setæ.

Habitat, 1 mile off Spanish Head, Isle of Man, in neritic material dredged from a depth of 16 fathoms; a number of specimens were obtained.

Remarks.—This species comes very near *Laophonte spinosa*, I. C. Thompson, especially in the structure of the antennules and mouth organs, but differs considerably in the structure of the second, third and fourth pairs of swimming feet; the outer branches of the second, third and fourth feet in *Laophonte spinosa* are two jointed and the inner three, whilst in *Pseudolaophonte aculeata* the second pair of feet consists of a single one-jointed branch, in the third pair each branch is composed of two joints and in the fourth pair the outer branch consists of three joints and the inner of two, the fifth feet also differ somewhat. The appendages of the male differ also from those of the male *Laophonte spinosa*.

Laophontodes bicornis, n. sp. Pl. III., figs. 24—25; IV., figs. 1—7.

Description of the species.—Female. Length ·5 millim. ($\frac{1}{25}$th of an inch). Body seen from above elongate narrow, the breadth gradually decreasing towards the posterior end; all the segments are more or less angular in shape and with the exception of the cephalic segment, bear each a **row** of short teeth on the distal margin. Cephalo-thoracic segment broadly triangular in outline, the frontal portion being produced into a small rostrum, and the lateral margins near the distal end into strong curved spines directed backwards and extending slightly beyond the middle of the second segment. Antennules short, five-jointed, all the joints are of moderate length except the fourth which is very short; the proportional lengths of the joints are as shown in the following formula :—

$$\frac{13 \cdot 17 \cdot 22 \cdot 3 \cdot 13}{1 \quad 2 \quad 3 \quad 4 \quad 5}$$

Antennæ small, two-jointed without any secondary appendage. Mandibles and other mouth organs nearly as in *Laophonte*. The first pair of swimming feet are similar to those of *Laophontodes typicus*, and the second, third and fourth pairs are also similar to the corresponding feet of that species. The fifth pair are large and prominent and project outwards from the sides of the fifth segment; each foot consists of a single narrow elongate branch, composed of two-joints, furnished with one seta on the inner distal angle of the first joint and two on the outer angle, the second joint has two setæ on the inner margin, two on the apex and one on the outer margin, the basal joint has also a row of cilia on its inner margin. Caudal stylets long and narrow, about equal to the combined lengths of the last two abdominal segments.

Habitat, Off Port Erin, from dredged material collected June, 1895; only one specimen has been observed.

Remarks.—This species is easily distinguished from *Laophontodes typicus* the only other member of the genus, by the lateral projections of the cephalothoracic segment, the proportional lengths of the joints of the antennules and the length of the caudal stylets·; the fifth feet also differ, in this species they are two-jointed whilst in *Laophontodes typicus* they are composed of a single joint only.

Normanella attenuata, n. sp. Pl. IV., figs. 8—20.

Description of the species.—Female. Length 1 millim. ($\frac{1}{25}$th of an inch). Body elongate cyclindrical, slender. Antennules nine-jointed; the second much longer than the others, seventh and eighth joints very small, the others are of moderate length as shown by the formula:—

$$\frac{9 \cdot 15 \cdot 10 \cdot 7 \cdot 4 \cdot 5 \cdot 1 \cdot 1 \cdot 5}{1 \quad 2 \quad 3 \quad 4 \quad 5 \quad 6 \quad 7 \quad 8 \quad 9}$$

Antennæ three-jointed, stout and of moderate length, a small one-jointed secondary branch arises from the lower distal end of the basal joint of the primary branch and is furnished with two setæ; the lower one of which appears to be articulated to the apex of the joint. Mandibles slender with a serrated apex, basal portion of the mandible palp considerably dilated and bearing two one-jointed branches, the outer branch being much longer than the inner. Maxillæ and foot-jaws nearly as in *Normanella dubia.* Inner branches of the first pair of swimming feet long and slender, two-jointed, basal joint longer than the entire outer branch, second joint about one-third the length of the basal joint, bearing one curved spine and two setæ on the apex, outer branches three-jointed, shorter than the basal joint of inner branches. In the second and third pairs of feet, the inner branches are short, and two-jointed; the outer branches are considerably longer than the inner and three-jointed. Inner branches of the fourth pair of feet three-jointed and very short, only reaching to about the middle of the second joint of the outer branches. Fifth pair of feet foliaceous, two branched, inner branch large and subtriangular bearing two setæ on the inner distal margin and two on the apex, outer branch pyriform, arising from the middle of the outer margin and extending considerably beyond the apex of the inner branch, bearing four setæ on its outer distal margin and two on the apex. Caudal stylets of moderate length, about twice as long as broad and fully half the length of the last abdominal segment.

Male. Antennules nine-jointed, sixth joint very short, the others of moderate length, hinged between the fourth and fifth joints and also between the seventh and eighth, all the other appendages with the exception of the fifth pair of feet are similar to the corresponding appendages of the female. The inner branch of the fifth pair sub-triangular in form bearing one stout plumose spine and two plumose setæ on its apex, the outer branch pyriform, bearing three setæ on its outer distal margin, and two on the apex, with a strong plumose spine between the two apical setæ.

Habitat, 1 mile off Spanish Head, Isle of Man, in neritic material dredged from a depth of 16 fathoms ; very few specimens were obtained.

Remarks.—This species differs considerably in shape from *Normanella dubia* but the structural details are almost similar to those on which the genus was founded, the only differences being that the antennules have nine joints instead of seven, and the inner branches of the fourth pair of feet have three joints instead of two. These differences are not considered to be of sufficient importance to warrant the establishment of a new genus for its reception.

Cletodes similis, T. Scott.

[T. Scott, Thirteenth An. Rep. Fish. Board. for Scot. Pt. III., p. 168, 1895.]

A few specimens were obtained from material washed

from sponges collected by Dr. Hanitsch at Port Erin, Isle of Man, in August, 1894. This species is very like *Cletodes lata* in general appearance but is easily distinguished from it on dissection by the structure of the antennules, the proportional lengths and armature of the outer and inner branches of the first pair of swimming feet, and also by the form of the fifth pair of feet.

Nannopus palustris, Brady.

Several specimens of this species were obtained in the mud collected from the Mussel beds near Duddon and from mud sent to the laboratory from the Fleetwood Oyster beds. It seems to be a brackish water species and in general appearance is very like *Platychelipus littoralis* another brackish water copepod, it can be distinguished from that species however, even without dissecting, by making an examination of the fifth pair of feet and also of the inner branches of the third and fourth pairs of feet. *Nannopus palustris* has two ovisacs and *Platychelipus littoralis* one only.

Idya elongata, n. sp. Pl. IV., figs. 21—24; Pl. V., figs. 1—5.

Description of the species.—Female. Length ·74 millim. ($\frac{1}{35}$th of an inch). Body seen from above elongate narrow, tapering rapidly towards the posterior end, the length being nearly equal to four times the greatest breadth; rostrum prominent with a bluntly rounded apex. Antennules short and comparatively stout; shorter than the cephalothoracic segment, eight-jointed; second and third joints longer than any of the others, as shown in the following formula :—

$$\frac{11 \cdot 16 \cdot 17 \cdot 13 \cdot 6 \cdot 8 \cdot 5 \cdot 12}{1 \quad 2 \quad 3 \quad 4 \quad 5 \quad 6 \quad 7 \quad 8}$$

Antennæ, mandibles and maxillæ nearly as in *Idya gracilis*, T. Scott. Foot-jaws also similar to those of that species but shorter and stouter. Inner branches of the first pair of swimming feet slender and of moderate length, basal joint nearly as long as the entire outer branch, and furnished with a plumose seta arising from the lower half of the inner margin and extending to slightly beyond the end of the branch, second joint fully two-thirds the length of the basal joint also furnished with a plumose seta arising from near the middle of its inner margin, third joint very small, bearing on its apex two stout spines and one short plumose seta; outer margins and proximal halves of the inner margins of the first and second joints fringed with short hairs, the joints of the outer branches are short and broad, the second joint is slightly shorter than the first and the third joint a little shorter than the second, the armature of the joints is somewhat similar to that of the first pair in *Idya furcata*; the spines are furnished with a row of moderately long cilia on the upper margins. Second, third and fourth pairs of swimming feet similar to those of *Idya furcata*. Fifth pair of feet very short being little more than half the length of the joint to which they are attached and extending only a little way beyond the base of the first segment of the abdomen, the length of each foot is about equal to twice the breadth, the secondary joint is furnished with three setæ on the apex, the innermost one being longer than either of the other two, outer very short; a short seta is attached to the outer margin a little way from the apex. Caudal stylets narrow and slightly divergent, length equal to about twice the breadth and nearly as long as the last segment of the abdomen.

Male. Antennules nine-jointed, hinged between the third and fourth joints and also between the seventh and eighth joints, fourth joint very small; the other appendages are similar to those of the female, fifth feet also similar to the fifth feet of the female but smaller.

Habitat, obtained from the mud collected on the Mussel beds between Morecambe and Heysham ; only a few specimens were obtained.

Remarks.—This species is very distinct from *Idya furcata* and also from two other species recently described— *Idya longicornis,* T. and A. Scott, and *Idya gracilis,* T. Scott—and can easily be recognised from either of them by the elongate form of the animal, the short antennules and the small fifth feet.

Idya gracilis, T. Scott.

[T. Scott, Thirteenth An. Rep. Fish. Board for Scot.; pt. III., p. 171, 1895.]

A number of specimens of this species were obtained from the shelly material dredged 1 mile off Spanish Head, Isle of Man, from a depth of 16 fathoms ; it is easily recognised by the long and slender inner branches of the first pair of swimming feet and also by the shape and arrangement of the setæ on the fifth pair of feet.

Family SAPPHIRINIDÆ, Thorell.

Modiolicola insignis, Aurivillius.

Living as a messmate within the mantle of the "horse mussel," *Mytilus modiolus.* A number of specimens were found in the examples of this Mollusc which were brought up in the trawl-net of the steamer, while working in the vicinity of the north end of "the Hole" on March 23rd, 1895. This appears to be a widely distributed species of Copepod, its range being probably co-extensive with that of the Mollusc. It has been recorded from the Firth of Forth, the Moray Firth, and from the vicinity of Mull. It has also been obtained in specimens of the same species of Mollusc dredged by Dr. Norman in 1893, off Trondhjem in Norway.

Family ASCOMYZONTIDÆ, Thorell (1859).

Dermatomyzon gibberum, T. and A. Scott.

[T. & A. Scott, An. & Mag. Nat. Hist., Ser. 6, Vol. XIII., p. 144, 1894.]

A considerable number of specimens of this species were obtained by washing the common starfish (*Asterias rubens*) in weak methylated spirit and afterwards examining the sediment. It was taken from starfish collected at Hilbre Island and afterwards from the same species of starfish taken in other parts of the district; both males and females were found, many of the latter with ovisacs attached.

Collocheres elegans, n. sp. Pl. V., figs. 6—15.

Description of the species.—*Female.* Length 1 millim. ($\frac{1}{25}$th of an inch). Body elongate, subpyriform, anterior segment large and somewhat triangular in outline and equal to twice the combined lengths of the second, third and fourth segments, rostrum small and inconspicuous. Antennules moderately long, slender and sparingly setiferous, twenty-jointed; the first, eighteenth and twentieth joints of about equal length and longer than any of the others, the second and tenth joints slightly smaller than the others; a sensory filament springs from the end of the third last joint. The following formula shows the proportional lengths of the joints :—

$$\frac{9 \,.\, 2 \,.\, 3 \,.\, 3 \,.\, 3 \,.\, 3 \,.\, 3 \,.\, 4 \,.\, 4 \,.\, 2 \,.\, 3 \,.\, 6 \,.\, 5 \,.\, 6 \,.\, 6 \,.\, 6 \,.\, 6 \,.\, 9 \,.\, 3 \,.\, 8}{1 \quad 2 \quad 3 \quad 4 \quad 5 \quad 6 \quad 7 \quad 8 \quad 9 \quad 10\,11\,12\,13\,14\,15\,16\,17\,18\,19\,20}$$

Antennæ three-jointed, basal joint long and narrow, bearing near the middle of the lower margin a small secondary branch, which consists of a single joint, nearly oval in outline and furnished with three small setæ on the apex and one near the middle of the upper margin, second joint of the antennæ about half the length of the first, third joint about two-thirds the length of the second and bearing at the apex a long slender spine having a slightly thickened base, and a small hair; a short seta also springs from near the base of the upper margin. Mandibles elongate narrow, denticulated on the oblique apex, palp rudimentary and consisting of a single moderately long hair. Maxillæ two-lobed, both lobes of about equal length, but one is slightly narrower than the other and is furnished with one seta at the apex, the broad lobe has four setæ on its apex. Foot-jaws somewhat similar to those of *Collocheres gracilicauda* (Brady). First four pairs of swimming feet also similar to those of that species; the outer branches of all the four pairs are armed with short dagger shaped spines and the terminal joint of the inner branch of the fourth pair is furnished with one stout dagger shaped spine on the apex and a smaller one near the middle of the outer margin. Fifth pair of feet somewhat rudimentary, two-jointed, basal joint broadly triangular in shape, the second joint which is attached to near the middle of the outer margin of the basal joint is elongate, curved, and bluntly serrated at its apex, the length being about equal to three and one-half times the breadth; it is furnished with three setæ, one on the apex and two a little lower down on the outer margin and slightly separated from each other. Abdomen slender, four-jointed, genital segment elongate narrow, length nearly equal to twice the breadth, and longer than the combined lengths of the next three segments, second joint about one-third the length of the first, third joint slightly smaller than the second, fourth joint smaller than the third. Caudal stylets about four times as long as broad and nearly equal to the length of the last two segments of the abdomen.

Habitat, off Port Erin, from dredged material collected June, 1895, only one specimen has been observed.

Remarks.—This species is not unlike *Collocheres gracili-*

cauda and may perhaps have been passed over for that species, but it can be readily distinguished from it by the much shorter caudal stylets and also by the shape of the fifth pair of feet.

Ascomyzon thompsoni, n. sp. Pl. V., figs. 16—26.

Description of the species.—Female. Length 1 millim. ($\frac{1}{25}$th of an inch). Body broad, suborbicular in shape, cephalothorax broadly ovate, last segment of thorax and abdomen much narrower, rostrum not prominent. Antennules slender, twenty-one-jointed, the first being the largest and ciliated on its upper margin; second to eighth joints small and of about equal length, ninth joint smaller than any of the others, eighteenth joint furnished with a short sensory filament. The proportional lengths of the joints are shown in the following formula :—

$$48.7.7.5.6.6.6.7.8.4.7.8.11.8.12.11.14.15.7.8.7$$
$$\overline{1\ 2\ 3\ 4\ 5\ 6\ 7\ 8\ 9\ 10\ 11\ 12\ 13\ 14\ 15\ 16\ 17\ 18\ 19\ 20\ 21}$$

Antennæ four-jointed, first joint long and bearing near the distal end of the lower margin, a small one-jointed secondary branch, which bears at the apex a moderately long seta, a small hair also springs from near the middle of the upper margin; second joint of the antennæ shorter and narrower than the first and having its lower margin ciliated, third joint very small, fourth joint about as long as broad and bearing at its apex one strong curved spine and two setæ. Mandibles slender, and stylet shaped; palp elongate narrow, two-jointed, second joint about one-third the length of the first and bearing at its apex, one long and one short plumose seta. The maxillæ consist of a short basal joint bearing two lobes of about equal length, but one is considerably narrower than the other, each lobe is furnished with four plumose setæ; one of the setæ on the broad lobe is much stouter and longer than the others, two of the other setæ on the same lobe are also comparatively stout but are only about half the length of the long seta. Anterior foot-jaws simple, bearing a strong curved apical claw. Posterior foot-jaws elongate slender, four-jointed, resembling those of *Dermatomyzon nigripes* (B. and R.). Both branches of the first four pairs of swimming feet short and stout, three-jointed and nearly equal in length. Fifth pair of feet rudimentary, two-jointed, inner joint short and broad, furnished with one plumose seta on its upper distal angle, outer joint elongate, length about equal to twice the breadth and bearing at its apex two moderately long plumose setæ and one small spine, both margins of the joint ciliated. Abdomen three-jointed, genital segment about as long as broad and nearly equal to the combined lengths of the next two segments and caudal stylets, second joint about half the length of the first, third joint about two-thirds the length of the second. Caudal stylets slightly longer than the last abdominal segment, length about equal to twice the breadth.

Habitat, 1 mile off Spanish Head, Isle of Man, in neritic material dredged from a depth of 16 fathoms; a few specimens only were obtained. A number of specimens have since been found in material washed from Ophiuroids (*Ophioglypha* and *Ophiothrix*) taken in the trawl-net off Blackpool, and sent to us by Mr. Ascroft.

Remarks.—This species is readily distinguished from the other members of the *Ascomyzontidæ* by the almost oval outline of the cephalothorax and on dissection by the structure of the mandible palp and maxillæ, the stout setæ on the larger lobe of the maxillæ appears to be a well marked character. Dr. W. Giesbrecht of the Zoological Station Naples, is preparing a monograph on this interesting family and an abstract which appeared in the Ann. and Mag. of Natural History for August, 1895, shows a number of changes in the nomenclature and classification of the genera and species.

SECTION V.

INVESTIGATIONS ON OYSTERS AND DISEASE.

(By Professor HERDMAN.)

FROM the earliest times more or less well grounded suspicion has been cast from time to time upon shellfish— chiefly oysters and mussels—as being the cause of outbreaks of disease amongst consumers. These outbreaks fall into two categories:—1st Cases of sudden poisoning due to the presence of putrefactive products, and 2nd Diseases due to a specific micro-organism, where there is a period of incubation and where therefore a considerable interval has elapsed between the infection and the actual illness. In the latter case it is obviously much more difficult to determine with certainty the source from which the disease germ has entered the body; and although many positive assertions have appeared of late years attributing outbreaks of enteric or typhoid fever to the consumption of oysters, still it must be pointed out that the connection between the two has not yet been scientifically proved, and is only at present more or less of a possibility or, at most, probability.

Under these circumstances I suggested to my colleague Professor R. Boyce that the subject was one well worthy of our attention, and during the past year we have been making a number of observations and experiments, both in our Liverpool laboratories and at Port Erin, upon the conditions under which oysters live healthily, and upon the possibility, or even in some cases the probability, of their being the carriers of disease germs. We gave a preliminary account of our work at the meeting of the British Association in Ipswich last September, and we shall prepare a detailed Report upon the matter to be laid before the next meeting of the British Association in Liverpool in September, 1896, but in the meantime so much public interest and apprehension has been raised by several recent outbreaks of typhoid popularly attributed to oysters, and the matter is so closely connected with the shellfish industries of this district, that I consider it advisable to give a summary here of the results of our work up to the present time.

A. The objects we had in view in entering on the investigation were as follows :—

1. To determine the conditions of life and health and growth of the oyster by keeping samples in sea waters of different composition—*e.g.*, it is a matter of discussion amongst practical ostreiculturists as to what specific gravity or salinity of water, and what amount of lime are best for the due proportionate growth of both shell and body.

2. To determine the effect of feeding oysters on various substances—both natural food, such as Diatoms, and artificial food, such as oatmeal. Here, again, there is a want of agreement at present as to the benefit or other-wise of feeding oysters in captivity.

3. To determine the effect of adding various impurities to the water in which the oysters are grown, and especially the effect of sewage in various quantities. It is known that oysters are sometimes grown or laid down for fattening purposes in water which is more or less con-taminated by sewage, but it is still an open question as to the resulting effect upon the oyster.

4. To determine whether oysters not infected with a pathogenic organism, but grown under insanitary conditions, have a deleterious effect when used as food by animals.

5. To determine the effect upon the oyster of infection with typhoid, both naturally—*i.e.*, by feeding with sewage water containing typhoid infection, and artificially—*i.e.*, by feeding on a culture in broth of the typhoid organism.

6. To determine the fate of the typhoid bacillus in the oyster—whether it is confined to the alimentary canal, and whether it increases in any special part or gives rise to any diseased conditions; how long it remains in the alimentary canal; whether it remains and grows in the pallial cavity, on the surface of the mantle and branchial folds; and whether it produces any altered condition of these parts that can be recognised by the eye on opening the oyster.

7. To determine whether an oyster can free its alimentary canal and pallial cavity from the typhoid organism when placed in a stream of clean sea water; and, if so, how long would be required, under average conditions, to render infected oysters practically harmless.

B. The methods which we employed in attaining these objects were as follows :—

I. Observations upon oysters laid down in the sea, at Port Erin—

(*a*) Sunk in 5 fathoms in the bay, in pure water.

(*b*) Deposited in shore pool, but in clean water.

(*c*) Laid down in three different spots in more or less close proximity to the main drain pipe, opening into the sea below low-water mark.

These observations were to ascertain differences of fattening, condition, mortality, and the acquisition of deleterious properties as the result of sewage contamination.

II. Observations upon oysters subjected to various abnormal conditions in the laboratory.*

(a) A series of oysters placed in sea water and allowed to stagnate, in order to determine the effect of non-aëration.

(b) Similar series in water kept periodically aërated.

(c) A series placed in sea water to which a quantity of fresh (tap) water was added daily, to determine effect of reduction of salinity.

(d) A series of oysters weighed approximately, and fed upon the following substances, viz. :—(1) Oatmeal, (2) Flour, (3) Sugar, (4) Broth, (5) Living Protophyta (Diatoms, Desmids, Algæ), (6) Living Protozoa (Infusoria, etc.), (7) Earth.

In this series of experiments the oysters were fed every morning and the water aërated, but not changed (evaporation was compensated for by the addition of a little tap water as required). The oysters were weighed from time to time, and observations made upon the apparently harmful or beneficial effects of the above methods of treatment.

(e) A series of oysters placed in sea water to which was added daily—

(1) Healthy fæcal matter.

(2) Typhoid fæcal matter.

(3) Pure cultivations of the typhoid bacillus.

The oysters were carefully examined to determine their condition, with special reference to condition of branchia, alimentary canal, adductor muscle, and viscera generally. The contents of the rectum, as well as the water in the pallial cavity, were subjected to bacteriological analysis

* The oysters were kept in basins in cool rooms of constant temperature, shaded from the sun, both at the Port Erin Biological Station and also in the Pathological and Zoological Laboratories at University College, Liverpool.

to determine the number of micro-organisms present, as well as the identity of the typhoid or other pathogenic organisms.

C. The following is a summary of the results obtained so far:—

We consider that these results are based upon tentative experiments, and serve only to indicate further and more definite lines of research. They must not be regarded as conclusive. We feel strongly that all the experiments must be repeated and extended in several directions.

Our experiments demonstrate:—

I. The beneficial effects of aëration—

(a) By the addition of air only;

(b) By change of water;

pointing to the conclusion that the laying down of oysters in localities where there is a good change of water, by tidal current or otherwise, should be beneficial.

II. The diverse results obtained by feeding upon various substances, amongst which the following may be noted. The exceedingly harmful action of sugar, which caused the oysters to decrease in weight and die; whilst the other substances detailed above enabled them to maintain their weight or increase. The oysters thrive best upon the living Protophyta and Protozoa. Those fed upon oatmeal and flour after a time sickened and eventually died.

III. The deleterious effects of stagnation, owing to the collection of excretory products, growth of micro-organisms, and formation of scums upon the surface of the water.

IV. The toleration of sewage, etc. It was found that oysters could, up to a certain point, render sewage-contaminated water clear, and that they could live for a prolonged period in water rendered completely opaque by the addition of fæcal matter; that the fæcal matter obtained from cases of typhoid was more inimical than that obtained from healthy subjects; and that there was considerable toleration to peptonised broth.

V. The infection of the oyster by the micro-organisms. The results of the bacteriological examination of the pallial cavity of the oyster, and of the contents of the rectum, showed that in the cases of those laid down in the open water of the bay the colonies present were especially small in number, whilst in those laid down in proximity to the drain pipe the number was enormous (e.g., 17,000 as against 10 in the former case). It was found that more organisms were present in the pallial cavity than in the rectum. In the case of the oysters grown in water infected with the *Bacillus typhosus*, it was found that there was no apparent increase of the organisms, but that they could still be identified in cultures taken from the water of the pallial cavity and rectum fourteen days after infection.

It is found that the typhoid bacillus will not flourish in clean sea water, and our experiments seem to show so far that it decreases in numbers in its passage along the alimentary canal of the oyster. It would seem possible, therefore, that by methods similar to those employed in the " Bassins de dégorgement " of the French ostreiculturist, where the oysters are carefully subjected to a natural process of cleaning, oysters previously contaminated with sewage could be freed of pathogenic organisms or their products without spoiling the oyster for the market.

It need scarcely be pointed out that if it becomes possible thus to cleanse infected or suspected oysters by a simple mode of treatment which will render them innocuous, a great boon will have been conferred upon both the oyster trade and the oyster-consuming public.

The discovery of a green tinge of more or less intensity which made its appearance on the mantle and other parts of the body of some of the oysters in our experiments started us on a series of investigations into the minute structure of the green parts, and the nature and causes of the greenness in general in oysters. We soon found that the greenness of our experimental oysters,* which is very different in appearance from the greenness of the culti-vated French oyster—the " huitre de Marennes "—was present also in some American oysters freshly imported, and was liable to make its appearance in oysters laid down in certain parts of our own coast. We found that this pale green inflammation, as it may be called, was known to the local oyster importers and oyster merchants, by whom the suggestion was made that the colour was due to copper poisoning in the oyster.

This led us to a chemical examination of the matter which showed (as had been shown in the case of other green oysters before) that copper had nothing to do with the disease. There was very little copper present in the green parts—practically no more than in the corresponding parts of colourless (yellow) oysters. Moreover we have kept oysters in old copper vessels and in vessels of sea water containing different amounts of sulphate of copper in solution (0·02 and 0·2 % of the copper, in 10 litres of water), and other salts, and in none of these cases did the animal acquire any green colour except what was deposited on the surface of the shell and other exposed parts until the death of the tissues when a certain amount of post-mortem green staining made its appearance.

Dr. Charles Kohn has kindly analysed the gills of some green (French) and some ordinary colourless oysters, and

* Our thanks are due to Charles Petrie, Esq,, C.C., and to George T. G. Musson, Esq., for the kind help they have given us in procuring various kinds of oysters for investigation and experiment.

finds that in both of them there are traces of copper, and of iron, but that the amount of both metals is actually greater in the colourless than in the green gills.

Cases of sudden poisoning following upon the consumption of oysters have frequently been ascribed to the oysters having a green colour which was supposed, with little or no reason, to be due to impregnation with a copper salt. The river Roach, in Essex, has long been known to produce in winter green oysters which, on account of their colour, are not sold in England but are sent to the French and other continental markets. It has been shown several times that copper has nothing to do with the greening of these oysters. However, that conclusion is constantly doubted, and such cases as the following are quoted:—In 1713 a certain Ambassador gave a great supper at the Hague, and, we are told, as a luxury procured green oysters from England. The guests who eat them are said to have been seized with severe colics and to have been cured with great difficulty. It is also said, however, that the merchant had palmed upon the Ambassador some common oysters tinted with copper instead of the true greens. Another historic case is that of the trial which took place at Rochefort in 1862 of a merchant who had sold green oysters imported from England and which were said to contain copper.

Other cases are on record of green oysters which are supposed to have taken up copper from old mines under the sea, as at Falmouth, or from the copper bottoms of ships, and so have become poisonous. Mr. G. C. Bourne of Oxford who has investigated the Falmouth oysters tells me that their greenness is in his opinion due to a green Desmid upon which the oysters feed in quantity. It is said that these oysters lose their colour on being transplanted to the mouth of the Thames.

It has been conclusively shown by various investigators, since Dumas in 1841 and Berthelot in 1855, that the green colour is not due to Chlorophyll and that although every oyster contains a very small amount of copper in its blood there is no reason to believe this becomes increased to a poisonous extent in any oyster, and that the green colour of the French cultivated oysters (" huitres de Marennes ") is not due to copper.

Gaillon in 1820 came to the conclusion that the green colour is due to the microscopic food which the oyster gets from the water of the " claire " or oyster park. He called the organism in question *Vibrio ostrearius*; it is now known to be a Diatom—*Navicula· fusiformis*, var. *ostrearia*, Grunow. Since then other French investigators, and notably Valenciennes in 1841, have corroborated Gaillon's conclusion and have shown that in addition to the branchiæ, the inner surfaces of the labial palps, the alimentary canal beyond the stomach, and to some extent the liver, become coloured green.

It has been proved experimentally by Puységur,[*] Bornet, Decaisne and others that white oysters can be greened rapidly—in 26 hours—by keeping them in clean soup plates and feeding them with water containing the *Navicula*. These interesting experiments were carried out at Le Croisic in Brittany, and were afterwards repeated in Paris; and the result seems entirely opposed to the old suggestion that iron salts in the soil at the bottom of the " claire " are the cause of the greening, which was alluded to twenty years ago by Bouchon-Brandely and has quite recently been revived by Carazzi at Spezia.

Ryder,[†] in America, was also in 1880 investigating the greening of oysters, with much the same results as those of Puységur. He went a step further, however, and

[*] Revue Maritime et Coloniale, 1880.

[†] U. S. Fish Comm. Report for 1882 (published 1884).

showed that the colouring matter was taken up by the amœboid blood cells, and that these wandering cells containing the pigment were to be found in the heart, in some of the blood-vessels and in aggregations in "cysts" under the surface epithelium of the body. He describes the colour (in the ventricle) as a "delicate pea-green," and states that it is not chlorophyll nor diatomine: he suggests that it may be phycocyanine or some allied substance.

In 1886, Ray Lankester* gave a useful summary of some of the earlier papers, and discussed the main questions concerned. Moreover he investigated the gills of the green oyster histologically, and described cells laden with green granules which occur in the epithelium of the gills and labial palps. He showed that such cells are also present in the common oyster, where, however, they are not green, and that these cells may be found also wandering over the surface of the gills. He considered them as "secretion" cells, but they are clearly the same structures which Ryder a few years before had found in the blood. Lankester found the Navicula in the intestine of the green oyster, and re-asserted that there was no copper and no iron in the refractory blue pigment—which he described under the name "marennin."

Quite recently, Chatin, de Bruyne, and others have re-investigated the structure of the oyster's gill and the process of greening in more detail, with the general result that the large cells (macroblasts) containing the green granules are now regarded as "phagocytes" conveying some substance to the surface of the body. One author, however, Carazzi, considers that the macroblasts are surface cells, which are taking up substances from without for purposes of nutrition, and he attributes the green

* Quart. Journ. Micr. Sci., Vol. XXVI., p. 71.

colour to sesquioxide of iron in the mud of the "claire." This revival of an old view I have alluded to above.

At present then there are several distinct views as to the exact cause and the meaning of the greenness of the French cultivated oyster which is fattened, flavoured and greened in the oyster parks or "claires" of La Tremblade, Marennes, Sable d'Olonne, Le Croisic and other places on the west coast of France. One view is that it is due to the microscopic food of the oyster, another that it is caused by the nature of the bottom of the "claire." One view is that it is a process of excretion or the removal of certain coloured matters from the body, another that it is a process of absorption of nutriment. The whole subject is at present under investigation both by ourselves and by others, and we hope to report upon our conclusions more fully in a few months. It cannot, however, be questioned that the normal green oyster of the French markets is in a thoroughly healthy condition and that its green colour is not due to copper nor to any other poisonous substance.

There are, however, other green oysters—or rather there is a greenness which may appear in any oyster under certain conditions—which have nothing to do with cultivation in "claires" and where the colour is not due to feeding upon diatoms. This is the pale greenness mentioned above which we have met with in some American oysters. We find that it is an inflammatory condition or "leucocytosis" in which enormous numbers of wandering leucocytes filled with large green granules come out on the surface of the body and especially on the mantle. For further details of this green condition of the oyster, with figures, I must refer to our full report to the British Association to be published later. We have tried growing oysters under various unusual conditions, including the addition to the sea water of fluid from alkali works, such as may enter our estuaries, in the hope of getting some clue to the cause of the green inflammation, but have so far failed to reproduce exactly in the laboratory the changes which seem to take place when the oyster is left in its natural (?) surroundings.

Our present opinion, however, is that oysters exhibiting this pale green leucocytosis are in an unhealthy state, and we may add that we find the liver in these specimens is histologically in an abnormal, shrunken and degenerate condition. Whether actually "unfit for food" or not, they are at any rate in very "poor" condition, and have lost the aroma and flavour of the normal healthy oyster.

It is clear that, so far as our present knowledge goes, oysters only share along with many other food matters— such as tinned foods, meat pies, fish under certain conditions, and diseased or infected meat—the responsibility of occasionally being capable of conveying poison, parasites, or disease germs into the human body; and that is, taken by itself, no sufficient reason why an important and highly esteemed food matter should be avoided. What is evidently necessary is that the precise conditions under which the oyster may become dangerous as human food should be investigated, and that when these are determined precautions should be taken to insure that the unhealthly conditions can never arise in our oyster beds.

It is all important that perfectly healthy grounds should be chosen for fattening the oysters upon. The water in which they are kept should be above suspicion. Oysters have to be fattened in water that is to some extent estuarine, and unfortunately estuaries are the places where a certain amount of sewage must find its way into the sea. When sewage is present in water the number of micro-organisms, pathogenic and otherwise, becomes very largely increased. The water of the Seine above Paris contains 300 organisms to the cubic centimetre, while below Paris the number has increased to 200,000 per cubic centimetre. We have also shown from our experiments at Port Erin that in oysters purposely placed near the outlet of the drain the number of organisms increased enormously. All estuaries, however, are not polluted, and the deleterious effects of sewage do not extend far, consequently there should be no great difficulty in finding perfectly suitable localities for oyster culture if a careful study is made of the matter. Tides and other currents, depth, specific gravity and temperature of the water all affect the distribution of the sewage in an estuary, and their influence should be carefully enquired into in connection with the site of any proposed shellfish cultivation.

Some of our experiments have shown that the oyster can purify polluted water in a most remarkable degree, but that property may have a bad as well as a good result. The good side of the matter is that the oyster in obtaining its nutriment from the water is able to convert useless and deleterious products of decomposition into excellent human food. The bad side of the matter is that if there happen to be any disease germs in the water the oyster may possibly strain out and store them up in its own body. And it even seems probable that other microbes associated with disease germs may play some part in causing or modifying disease.

It is re-assuring, however, to find as we do from our own investigations as well as from the consideration of the work of others that the typhoid organism (*Bacillus typhosus*) dies off very rapidly in ordinary sea water as one passes either in distance or time from the source of supply.

Professor Boyce has supplied me with the following facts in regard to the presence and behaviour of the typhoid Bacillus in sea water.

We showed at the British Association Meeting in 1895 that, as was to be expected, the number of micro-organisms in oysters grown in the vicinity of sewage was enormously increased. The organisms present were non-pathogenic for man, and it became necessary therefore for us to investigate the growth of the typhoid bacillus in sea water. The first point of importance is the relative proportion of typhoid sewage to ordinary sewage. The proportion of typhoid fæcal matter which may find its way into the sewers has been investigated by Laws and Andrewes, and in one instance in London they gave the proportion as $\frac{1}{250.000}$, pointing out, however, that in the case of the fever hospitals every endeavour was made to disinfect the typhoid materials. The same authors were only able to demonstrate the presence of the *Bacillus typhosus* in the drains in the immediate vicinity of the Typhoid Hospital. Further, from their experiments there seems every reason to suppose that sewage is an unfavourable medium for the propagation of the typhoid Bacillus, and that although when *incubated* and grown in *sterile* sewage the organism may show a slight multiplication in the first 24 hours, it soon tends to become extinct.

Since 1886 a very large number of investigators have shown that drinking and river water may become infected with typhoid sewage, and here again numbers of experiments have been made to ascertain whether the *Bacillus typhosus* propagates in fresh water. Kraus showed that a very rapid decline of the Bacillus and a very rapid increase of the ordinary water bacteria took place when the water was incubated. The most recent observations are those by Frankland and Ward, and they showed that the Bacillus disappeared at the end of 34 days in unsterilised Thames water and that there was no multiplication in potable water. Observations thus tend to show that neither sewage nor fresh water are favourable media and that the former is the least favourable.

A further very important point is the action of salt water upon the typhoid bacillus. In 1889 Giaxa made observations upon the vitality of the B. typhosus in sterilised and unsterilised sea water and showed that it was present in the latter up to the 9th day, and in the former to the 25th. Frankland and Ward showed that a 3 % salt solution most prejudicially influences the growth of the Bacillus, the latter disappearing by the 18th day.

Our own experiments have been made so far with sterilised sea water incubated at 35°C., and in one case at 8°—10°C· A culture of the B. typhosus on agar was emulsified with sterilised water and a definite quantity of this in each instance was added to the sterilised sea water.

Experiment I. No. of bacilli at time of mixing 29,250.

After	21 hours	....	20,475.
,,	45 ,,	....	9,945.
,,	71 ,,	....	9,360.
,,	95 ,,	....	5,850.
,,	271 ,,	....	260.
,,	340 ,,	....	11.

Experiment II. At time of mixing .... 1,300.

After	21 hours	....	1,105.
,,	45 ,,	....	780.
,,	71 ,,	....	650.
,,	95 ,,	....	325.
,,	271 ,,	....	2.
,,	340 ,,	....	0.

Experiment III. At time of mixing.............. 22,750.

After 5 hours .... ·17,550.

After 23 hours .... 11,700.

,,   48   ,, ... 3,250.

,,   72   ,, ... 3,250.

,, 247   ,, ... 455.

,, 316   ,, ... 325.

Experiment IV. At time of mixing .... 130.

After 5 hours .... 41.

,, 23   ,, ... 31.

,, 48   ,, ... 38.

,, 72   ,, ... negative.

,, 247   ,, ... 1.

,, 316   ,, ... 0.

Experiment V. At time of mixing .... 31,200.

After 172 hours .... 9,360.

,, 244   ,, ... 325.

Experiment VI. At time of mixing .... 325.

After 172 hours .... 2.

Experiment VII. At time of mixing .... 32,500.

After 504 hours (water kept at 8°C. to 10°C.). 79.

Experiment VIII. At time of mixing .... 325.

After 504 hours .... 0.

These results are fairly uniform. When a large number of Bacilli are added to the water their presence may be demonstrated longer than in cases where smaller quantities are used. Fourteen days would appear to be the average duration in sea water incubated at 35°C., whilst when kept in the cold their presence was demonstrated on the twenty-first day. There appears to be no initial or subsequent multiplication of the Bacilli. Between 40 and 70 hours after infection there is less decrease than at other periods; but there is no evidence of increase in numbers of the Bacilli when grown in sea waters either when incubated or at ordinary temperatures.

- On the whole the investigations which are summarized

above—and which it must be remembered are not yet finished—give results of a re-assuring nature, and demand from the public at the very least a suspension of judgment, while they also indicate the advantage of adopting some simple sanitary measures which, if properly carried into effect, would go far to remove suspicion from the oyster in our markets. These measures are, 1° a strict examination of all grounds upon which oysters are grown or bedded so as to ensure their freedom from sewage, and 2°, if practicable, the use of " dégorgeoirs " or disgorging tanks in which the oysters should be placed for a short time before they are sent to the consumer.

EXPLANATION OF THE PLATES.

PLATE I.

Stenhelia herdmani, n. sp. (A. Scott).

Fig. 1. Female seen from the side, × 27.   2. Antennule, × 63.   3. Antenna, × 85.   4. Mandible, × 85. 5. Maxilla, × 85.   6. Anterior foot-jaw, × 127. 7. Posterior foot-jaw, × 90.   8. Foot of first pair of swimming feet, × 85.   9. Foot of fourth pair, × 85.   10. Foot of fifth pair, × 127.   11. Abdomen and caudal stylets, × 170.

Stenhelia similis, n. sp.

Fig. 12. Female seen from the side, × 40.   13. Antennule, × 127.   14. Antennule of male, × 127. 15. Antenna, × 125.   16. Mandible, × 253. 17. Maxilla, × 253.   18. Posterior foot-jaw, × 253.   19. Rostrum, × 253.   20. Foot of first pair of swimming feet, × 127.   21. Foot of fourth pair, × 127. 22. Foot of second pair, male, × 127. 23. Foot of fifth pair, × 125. 24. Foot of fifth pair, male, × 125. 25. Abdomen and caudal stylets, × 53.

PLATE II.

Ameira gracile, n. sp.

Fig. 1. Female seen from the side, × 64. 2. Antennule, × 152. 3. Antennule, male, × 152. 4. Antenna, × 253. 5. Mandible, × 380. 6. Posterior foot-jaw, × 380. 7. Foot of first pair of swimming feet, × 170. 8. Foot of fourth pair, × 170. 9. Foot of fifth pair, × 380. 10. Foot of fifth pair, male, × 380. 11. Abdomen and caudal stylets, × 80.

Canthocamptus palustris, Brady.

Fig. 12. Female seen from the side, × 50. 13. Antennule, × 200. 14. Antennule, male, × 152. 15. Antenna, × 253. 16. Mandible, × 300. 17. Posterior foot-jaw, × 380. 18. Foot of first pair of swimming feet, × 170. 19. Foot of fourth pair, × 170. 20. Foot of fifth pair, × 253. 21. Foot of fifth pair, male, × 253. 22. Appendage to the first abdominal segment, male, × 253. 23. Abdomen and caudal stylets, × 80.

Tetragoniceps trispinosus, n. sp.

Fig. 24. Posterior foot-jaw, × 300. 25. Foot of fifth pair, × 380.

PLATE III.

Tetragoniceps trispinosus, n. sp.

Fig. 1. Female seen from above, × 80. 2. Antennule, × 170. 3. Antenna, × 253. 4. Foot of first pair of swimming feet, × 253. 5. Foot of fourth pair, × 253. 6. Abdomen and caudal stylets, × 253.

Pseudolaophonte aculeata, n. gen. and n. sp.

Fig. 7. Female seen from above, × 106. 8. Antennule, × 170. 9. Antennule, male, × 170. 10. Antenna, × 125. 11. Mandible, × 253. 12. Maxilla, × 253. 13. Anterior foot-jaw, × 253. 14. Posterior foot-jaw, × 253. 15. Foot of first pair of swimming feet, × 253. 16. Foot of second pair, × 253. 17. Foot of third pair, × 253. 18. Foot of fourth pair, × 253. 19. Foot of fifth pair, × 170. 20. Foot of third pair, male, × 253. 21. Foot of fourth pair, male, × 253. 22. Foot of fifth pair, male, × 380. 23. Abdomen and caudal stylets, × 80.

Laophontodes bicornis, n. sp.

Fig. 24. Mandible, × 500. 25. Anterior foot-jaw, × 500.

PLATE IV.

Laophontodes bicornis, n. sp.

Fig. 1. Female seen from above, × 120. 2. Antennule, × 253. 3. Antenna, × 253. 4. Posterior foot-jaw, × 380. 5. Foot of first pair of swimming feet, × 253. 6. Foot of fourth pair, × 253. 7. Foot of fifth pair, × 253.

Normanella attenuata, n. sp.

Fig. 8. Female seen from the side, × 50. 9. Antennule, × 127. 10. Antennule, Male, × 127. 11. Antenna, × 150. 12. Mandible, × 253. 13. Maxilla, × 253. 14. Posterior foot-jaw, × 253. 15. Foot of first pair of swimming feet, × 150. 16. Foot of second pair, × 150. 17. Foot of fourth pair, × 150. 18. Foot of fifth pair, ×

300. 19. Foot of fifth pair, male, × 300. 20. Abdomen and caudal stylets, × 90.

Idya elongata, n. sp.

Fig. 21. Posterior foot-jaw, × 380. 22. Foot of fifth pair, female, × 115. 23. Foot of fifth pair, male, × 115. 24. Appendage to the first abdominal segment, × 115.

PLATE V.

Idya elongata, n. sp.

Fig. 1. Female seen from above, × 64. 2. Antennule, × 253. 3. Antennule, male, × 253. 4. Foot first pair of swimming feet, × 190. 5. Foot fourth pair, × 190.

Collocheres elegans, n. sp.

Fig. 6. Female seen from above, × 52. 7. Antennule, × 133. 8. Antenna, × 170. 9. Mandible, × 253. 10. Maxilla, × 170. 11. Anterior foot-jaw, × 170. 12. Posterior foot-jaw, × 170. 13. Foot of first pair of swimming feet, × 125. 14. Foot of fourth pair, × 125. 15. Foot of fifth pair, × 190.

Ascomyzon thompsoni, n. sp.

Fig. 16. Female seen from above, × 50. 17. Antennule, × 133. 18. Antenna, × 190. 19. Mandible, × 190. 20. Maxilla, × 125. 21. Anterior foot-jaw, × 127. 22. Posterior foot-jaw, × 127. 23. Foot of first pair of swimming feet, × 127. 24. Foot of fourth pair, × 127. 25. Foot of fifth pair, × 200. 26. Abdomen and caudal stylets, × 85.

NOTES on the DISTRIBUTION of AMPHIPODA—
The proportion of species to genera in small and large
areas; also of species at various depths.

By Alfred O. Walker, F.L.S.

[Read March 13th, 1896.]

The remarks of Prof. Herdman in his Presidential
Address to Section D (Zoology) at the meeting of the
British Association last September (1) as to the relative
proportion of genera and species in large and small areas,
and (2) on Dr. Murray's theory that the number of
species of marine animals is greater, per haul, at and
near the 100 fath. line than in depths less than 50 fath.,
have led me to make some investigations into the
distribution of the Amphipodous Crustacea in our district
as compared with Norway.

I have chosen the Amphipoda in preference to any
other order of Crustacea because those of Norway have
recently been described by Prof. G. O. Sars in the first
part of his magnificent work on the Crustacea of Norway,
and also because more than half of the species of
Malacostraca in our district belong to it.

The proportions of genera to species have already been
given by Dr. Herdman in the Ninth Report of the
L.M.B.C. (1896), and I will only repeat the summary:

Rhos Bay, 1 to 10 fath., species are to genera as
115:100.

L.M.B.C. District, shore to 70 fath., species are to
genera as 159:100.

Norway, shore to 1215 fath., species are to genera as
232:100.

Now as to the second question, *viz.*, the proportion of species at different depths, I find on taking out all the species of Sars of which the depth at which they occur is given (but omitting pelagic and parasitic species) the results shown in the annexed summary.

And, as it is impossible for species to be taken at a greater depth than the dredge has gone, while it is not only possible, but very probable, that the dredge will catch some specimens between the bottom and the surface while being hauled up, it is more likely that species will be recorded at too great a depth than the reverse.

Where a range of depth is given I have taken the mean, *e.g.*, 20 to 100 fath. is reckoned as 60 fath.

The following L.M.B.C. species have not yet been found in Norwegian waters, those marked " M." being Mediterranean species:

Lysianax longicornis (Lucas). M.

Socarnes erythrophthalmus, Robertson.

Nannonyx spinimanus, Walker.

Phoxocephalus fultoni, Scott. M.

Amphilochus melanops, Walker.

Cyproidea brevirostris, Scott.

Syrrhoe fimbriatus, Stebbing and Robertson.

Guernea coalita (Norman).

Mæra batei, Norman.

Leptocheirus pilosus, Zaddach. M.

Photis pollex, Walker.

Podocerus ocius, Bate. M.

P. cumbrensis, Stebbing and Robertson.

Unciola crenatipalmata (Bate).

Colomastix pusilla, Grube. M.

The following is a summary of the depths at which the

Amphipoda Gammarida and Caprellida occur according to Sars and in the L.M.B.C. District:—

Number of Sars' species under   20 fath. (A)   106.

    do.   20— 50 ,, (B)   102.

    do.   50—100 ,, (C)   74.

    do.   over 100 ,, (D)   57.

Number of Sars' species in L.M.B.C. Dist. (A)   76

,,   species not in Sars in   do.   14

    — 90.

,,   Sars' species, &c.   ...   (B)   21

,,   species not in Sars   ...   2

    — 23.

,,   Sars' species, &c.   ...   (C)   9

,,   species not in Sars   ...   0

    — 9.

,,   Sars' species, &c.   ...   (D)   — 1*.

It will be seen from the above, at least as regards the Amphipoda,

1. That the proportion of species to genera increases with the area, and

2. That a far greater number of species are found at depths less than 100 fath. than at that or a greater depth, 208 species being recorded below 50 fath. against 74 between 50 and 100 fath. and 57 above 100 fath.

---

* *Podocerus odontonyx* (= *P. herdmani*), a doubtful species taken in 21 fath. and 3? fath. off Port Erin.

FERNS and FLOWERING PLANTS: a CHAPTER in EVOLUTION.

AN ADDRESS BY

D. H. Scott, M.A., Ph.D., F.R.S.,

HONORARY KEEPER OF THE JODRELL LABORATORY, ROYAL GARDENS, KEW.

[Read May 8th, 1896.]

ABSTRACT.

THE object of the lecture was to show how much light has been thrown on the evolution of certain groups of Flowering Plants, by modern work on fossil botany.

The order Cycadeæ, now represented only by 9 genera with about 70 species appeared to be the remnant of what was once a great class of plants, with wide-spreading affinities.

The first part of the lecture was devoted to plants from the Palæozoic rocks, which combine the characters of Ferns with those of Cycads. *Lyginodendron*, a Coal-measure fossil the structure of which was discovered by the late Prof. Williamson, was shown to have a stem which was essentially that of a Cycad, while the leaves, both in form and structure, were entirely Fern-like, belonging to Brongniart's genus *Sphenopteris*. The adventitious roots, when young resembled those of Marattiaceæ, but subsequently grew in thickness like the roots of modern Gymnosperms. *Heterangium* was also described, and proved to stand nearer to the Ferns (resembling those of simple structure, such as *Gleichenia*), as might be expected from its greater geological age, one species appearing in the Burnt island beds, at the base of the Carboniferous formation.

Other genera, especially *Protopitys* and *Medullosa*, were briefly referred to, as also showing a combination of Fern-like with Cycadean characters, and indicating that there were probably several distinct lines of descent, starting from various groups of the Ferns, and leading in the direction of Cycad-like plants. The evidence, though at present almost limited to vegetative characters, was regarded as sufficient to establish the existence of an extensive intermediate group in the borderland between Ferns and Cycadeæ, taking the latter term in a wide sense. The lecturer, however, pointed out that these intermediate forms must not be regarded as belonging to the direct line of descent of our living Cycads; the genealogical tree of the Vegetable Kingdom (like that of animals) was far more complex than we easily realized, and the chances were much against our lighting on "missing links," *i.e.*, the direct ancestors of living plants.

It now seemed certain that a number of the well-known Fern-like fronds of the Coal-measures, including species of the so-called genera *Sphenopteris*, *Alethopteris*, and *Neuropteris*, really belonged to these transitional forms, though many others were known to be true Ferns.

Passing on to the evidence from the Secondary rocks, the lecturer dwelt on the abundance and beauty of the Cycadean remains representing both leaves and stems from the Oolite, Wealden and Lower Greensand. He pointed out, however, that only in the rarest cases was there any evidence for the fructification having been of a truly Cycadean type. The best known specimens of reproductive organs in these plants were of quite peculiar structure, totally unlike the cones of any existing Cycads. The classical genus *Bennettites*, which might be called the *Archæopteryx* of the Vegetable Kingdom, was described at length. These remarkable plants, now so well known from the researches of Carruthers, Solms-Laubach and others, strongly resembled Cycadeæ in habit and in their vegetative anatomy ; their fructifications, however, were of a highly complex character, consisting of a receptacle bearing numerous seeds on long pedicels, with intermediate scales, the whole being enclosed in a kind of pericarp, and surrounded by bracts. The seeds were perfectly preserved, and were found to be exalbuminous (unlike those of any Gymnosperm) and to contain a large Dicotyledonous embryo. The fructification thus approached that of an Angiosperm, but at the same time *Bennettites* and its allies retained Fern-like characters in their ramenta, or scale-hairs, which were identical with those so common in Ferns, and quite different from anything in existing Cycadeæ. The facts indicated that the Bennettiteæ (which were probably an extensive family) though possessing clear Cycadean affinities, were not on the same line of descent with recent Cycads, but rather indicated a "short cut" from Filicineæ towards Angiosperms. Here again a warning was necessary against the assumption of direct ancestry. While *Bennettites* certainly showed an approach to Angiospermous organization, it was probably not directly related to any of the existing groups of Angiosperms.

It appeared then that the existing Cycadeæ represented a surviving offshoot of what was once an important stock, connecting the Filicineæ with various groups of Flowering Plants. Only a small part of the question, however, could be dealt with in the time available.

The lecture was illustrated by between 40 and 50 lantern-slides, representing existing Cycadeæ, and the structure of the various fossil plants described. Some diagrams, the work of the late distinguished Palæobotanist, Prof. W. C. Williamson of the Owens College, were also exhibited, as well as a number of specimens from the botanical museum of the University College, Liverpool.

Before the lecture a demonstration of microscopic slides, showing the structure of the fossil plants in question, was held in the botanical laboratory.

ADDITIONAL OBSERVATIONS on the VITALITY and GERMINATION of SEEDS.

By A. J. Ewart, B.Sc., Ph.D. (Leipzig).

1851 EXHIBITION SCHOLAR.   LATE DEMONSTRATOR·OF BOTANY IN
UNIVERSITY COLLEGE, LIVERPOOL.

In the present paper it is proposed to give an account of additional experiments which have been performed, illustrating and corroborating several of the points discussed in a previous paper.* The results therein arrived at, in three directions merited further investigation. These are; the effects of prolonged immersal in water; the resistance to dessication, and the oxytropic irritability of young radicles.

The results of immersal in water in which the development of Bacteria, Infusoria, etc., is allowed to continue unchecked (temperature 15—20°C.) are shown in Table A., the numbers given being comparison percentages with normal seeds.

TABLE A.

	5 days.	10 days.	14 days.	3 weeks.	4 weeks.
Peas - -	66 p.c. germinated.	9 p.c.	None.		
Haricots -	90 p.c. ,,	None.		None.	
Barley -	64 p.c. ,,	18 p.c.	6 p.c.	None.	
Hemp - -	85 p.c. ,,	60 p.c.	22 p.c.	19 p.c.	15 p.c.

* Trans. L'pool Biol. Soc., Vol. VIII., 1894.

Under conditions as above the seeds may commence to germinate and it is also impossible to be certain whether or not the apparent injurious influence of prolonged soaking in water is not merely due to the development of Bacteria and the putrefactive changes thereby induced. To eliminate these factors the following precautions were taken.

The dry seeds are well washed with a watery solution of mercuric chloride and then with boiled sterilized water, in a tube of which they are finally left. The tube is completely filled with water and the mouth is hermetically sealed with a plug of cotton wool soaked in melted paraffin or wax. With Peas, Haricots and Barley, owing to the absence of all oxygen from the surrounding water no germination whatever takes place, but with Hemp, probably owing to the fact that the seeds themselves may contain a little air, *i.e.*, oxygen, in a good many cases the seed coat splits, and the radicle protrudes and may elongate a millimetre or two.

The mode of sterilization adopted is not always successful but is the only one practicable without damaging the seeds and with sufficient care and accuracy of manipulation gives quite good results. During the experiments the seeds are exposed to a temperature varying from 4—10°C. The results obtained are shown in Table B.

TABLE B.

Seeds in sterilized oxygenless water for :

	1 week.	2 weeks.	3 weeks.	4 weeks.	14 weeks.
Peas - -	75 p.c.	12 p.c.	3 p.c.	None.	
Haricots -	12 p.c.	Few seeds cotyledons living. None germinate.	None.	.	

	1 week.	2 weeks.	6 weeks.	10 weeks.	14 weeks.
Barley -	88 p.c.	53 p.c.	36 p.c.	26 p.c.	18 p.c.
Hemp - -	99 p.c.	93 p.c.	48 p.c.	11 p.c.	9 p.c.

Of the Haricots planted after two weeks immersal none germinate but in a few seeds the cotyledons are living and may remain living for a week or two. It appears that a comparatively short immersion in water is fatal to seeds containing much proteid food material, whilst starchy and oily seeds, such as Barley and Hemp, may show marked resistant powers. The apparently lower resistant power of the Hemp, as compared with the Barley, is really due to the fact that a large number of the seeds undergo incipient germination. Thus, of the Hemp seeds taken after ten weeks immersal in sterilized water, in 60 p.c. the seed coat is split and the tip of the radicle just protruding. Of these 8 p.c. germinate. In 40 p.c. the seed coats are quite entire and 7 p.c. germinate. In a normal sample of these Hemp seeds only 75 p.c., however, germinate and all the seeds which were dead to commence with are included amongst those which show no signs of germination. Hence, of the living seeds with entire integuments, the power of germination is retained by 20 p.c. after ten weeks immersal but of those which undergo incipient germination only by 8 p.c.

It follows from the above experiments that in certain cases seeds can, in the absence of oxygen, though soaked with water, remain in a dormant condition and retain their vitality for long periods of time. If germination has begun the seeds are very much less resistant, young seedlings being readily asphyxiated by the absence of oxygen.

Further experiments were also performed upon the resistance of seeds to dessication. Table C. gives the results of a series of such experiments the seeds being dried over sulphuric acid at from 37°C.—38°C. for the times given. The numbers given are comparison percentages with seeds kept for similar lengths of time in paper bags at a room temperature from 16°C.—20°C.

TABLE C.

	Cucurbita pepo.	Brassica napus	Helianthus annuus.	Barley.	Hemp.	Peas.	Haricots.
3 weeks.	73 p.c.	48 p.c.	83 p.c.	98 p.c.	None.	15 p.c.	None.
6 weeks.	50 p.c.	43 p.c.	51 p.c.	57 p.c.	None.	None.	None.

Seeds of *Brassica napus* after nine weeks dessication gave as their comparison percentage 21 p.c. but of these many seedlings are weaklings and soon die. Of the seeds of *Helianthus* which germinated after six weeks drying, in 9 p.c. the radicle was killed, the " seedlings " consisting of hypocotyledonary axis, cotyledons and plumule, all soon dying. Similarly in 5 p.c. of the pea embryoes formed after three weeks drying no radicle or roots develop, the plumule and cotyledons soon dying and decaying, but in some cases not till the plumule has attained a length of several centimetres ; in 6 p.c. the radicle does not develop but is replaced by secondary roots ; the remaining 4 p.c. are normal seedlings but in a few cases the stems are irregularly twisted and grow horizontally instead of upright. In this last case apparently one result of dessication has been to induce irregularities of growth in the plumule, so pronounced in character, as to inhibit or mask its ageotropic and heliotropic irritabilities. With shorter periods of dessication Haricots and Hemp gave the following results.

TABLE D.

	1 week.	10 days.	2 weeks.	3 weeks.
Haricots - -	39 p.c.	15 p.c.	3 p.c.	None.
Hemp - - -	52 p.c.	35 p.c.	7 p.c.	None.

In both cases several of the young seedlings are weaklings and soon die, so that the percentage of seeds capable of forming normal healthy adult plants is still less than that of those which remain capable of germination. Prolonged dessication at a moderately raised temperature therefore exercises a marked injurious influence upon all seeds, one sign of which is the increase, which is frequently very pronounced, in the time taken in germinating by those seeds which remain living and capable of germination. The resistant power of a seed to dessication is partly dependent upon the nature and thickness of the seed coats and appears also to be connected with the form in which the reserve food material is stored. Thus, other conditions being similar albuminous seeds appear to be the least resistant to dessication, oily seeds next, and starchy seeds most resistant. At ordinary room temperatures (15°C.—20°C.) dessication is always more or less imperfect and under such conditions oily seeds appear to be quite as resistant as starchy ones (pp. 229 and 230 of previous paper). In very many cases seeds are very intolerant of even ordinary air drying. Thus seeds of *Oxalis*, *Salix* and *Populus* are unable to withstand from one to three weeks air drying, whilst prolonged air drying also exercises a marked injurious influence upon the seeds of *Acer*, *Fagus*, *Aesculus*, etc.*

* Schröder. Ueber die Austrocknungfähigkeit der Pflanzen. Bot. Untersuch. Tübingen Bd. II., Hft. 1, 1886.

The most resistant of seeds, however, cannot have the percentage of water which they contain reduced to lower than between 2 and 3 p.c. of their dry weight without their vitality being affected. It is possible that this water forms an integral part of the dried stable protoplasmic molecule and in this connection it is interesting to notice that Loew's[*] equations for the production of proteid by the polymerization of aspartic aldehyde and addition of hydrogen and sulphur give as the final result an albumen having the formula

$$C_{72} H_{112} N_{18} SO_{22} + 2 H_2O, \text{ or } C_{72} H_{112} N_{18} SO_{22} 2 H_2O,$$

and containing 2·2 p.c. of water. It appears possible that this or some analogous empirical formula may represent the chemical composition of protoplasm when in the dried condition. A seed is resistant to dessication when its protoplasm can assume and maintain this particular composition for long periods of time. It is interesting to notice that no cell the protoplasm of which shows streaming or rotation can withstand dessication. It appears that where the vital activity of the plasma manifests itself in the form of rotation or circulation, the preservation of vitality is indissolubly connected with the presence of free fluid water in the protoplasm.

OXYTROPIC IRRITABILITY OF ROOTS.

In the previous paper an account was given of certain experiments which seemed to lead to the conclusion that the root apices of plants possess a special oxytropic irritability enabling them to seek out the regions in which oxygen is most abundant and avoid those in which it is deficient or absent. Satisfactory demonstration of the presence of such an irritability must always, owing to the

[*] Loew. Die Chemische Kraftquelle in lebenden Protoplasma. München, 1882.

nature and varying character of the stimulus and to the extreme difficulty of successfully controlling its application, necessarily be extremely difficult.

If Hemp and Peas are placed in a tube filled with boiled sterilized water and closed at the open end by a loose plug of cotton wool it is commonly found that the radicles of the seeds immediately, in contact with the plug of cotton wool, grow upwards against the action of gravity and finally emerge from the open end of the tube, then growing vertically downwards. If the tubes are kept slanted, at a small angle with the horizontal so as to diminish the antagonism between the geotropic and oxytropic irritabilities of the radicles, the upward curvatures of these towards the source of oxygen and away from parts where it is absent or deficient, are more markedly produced. The effect might possibly partly be due to somatotropism the radicles preferring rather to burrow in the damp cotton wool than to immerse themselves in the subjacent water. This error is, however, easily obviated by filling the lower parts of the tubes with cotton wool soaked in boiled sterilized and practically oxygenless water. Above this the soaked seeds are placed and then covered by a loose plug of cotton wool. For purposes of comparison similar experiments are also made with tubes open at both ends. Both sets of tubes are placed at a slight angle 15°— 30°, with the horizontal and kept in a moist well aerated atmosphere at a temperature of 25°C. In the tubes, open at both ends, the radicles unless affected by somatotropism grow directly downwards, but in the tubes in which only the upper end is open in most cases the radicles bend upwards and may go on growing upwards against the action of gravity until they reach the mouth of the tube when they at once grow downwards. In many cases the radicle after bending upwards for a short distance and penetrating the upper plug of cotton wool, then bends downwards again and this may be repeated more than once. This upward curvature, followed by a downward one, is due to the fact that the radicle at first bends upwards in virtue of its oxytropic irritability until having penetrated the upper plug of cotton wool, it reaches and is surrounded on all sides by an atmosphere rich in oxygen, when in virtue of its geotropic irritability it curves downwards again. In a few cases the radicles grow downwards into the lower cotton wool for a short distance, growth becoming slower and slower as the radicle penetrates deeper and deeper. In such cases the tip of the radicle was probably equally surrounded on all sides by an atmosphere poor in oxygen and rich in CO_2 and hence being equally affected on all sides its oxytropic irritability or tendency to bend towards regions richer in oxygen was not called into play. Experiments performed with a variety of seeds gave in general corroboratory results and though the oxytropic irritability is not always equally well marked, nevertheless it can generally be made under appropriate conditions to overcome both geotropism and somatropism.

In *Helianthus annuus* the primary root shows marked oxytropism, which in the secondary roots is only slight and is weaker than their somatropism. *Pisum* is similar to *Helianthus* but the oxytropism of the secondary roots is rather more pronounced. In *Cucurbita* the radicles show distinct oxytropism and but slight somatropism but in the secondary roots the somatropism is marked and the oxytropism only slight. The primary radicles of Hemp also show oxytropism but their somatropism is very strong and in the secondary roots quite overpowers the other directive agencies.

That roots develop and flourish more abundantly in

oxygenated areas than in ones deficient in oxygen is well known, and it appears that the primary radicle more especially and to a lesser extent the secondary roots possess the power of seeking out within given limits and under appropriate conditions of stimulation the better oxygenated areas of soil and turning to them in preference to areas deficient in oxygen.

Since the above was written I have found that H. Molisch (Ueber die Ablenkung der Wurzehn von ihrer normalen Wachsthumsrichtung durch Gase. Ber. D. Bot. Gesell., 1884, II., p. 160) has already described this special form of irritability as occurring in the roots of *Zea*, *Cucurbita* and P*isum*, and has given it the name of "aerotropismus." The term "oxytropismus" seems to me to be better than "aerotropismus," but the latter has of course priority. The experiments here described support in all essentials the conclusions drawn by Molisch. The curvatures, however, which he found could be induced in many roots by the action of various poisonous gases are probably to be regarded as Traumotropic in nature and do not seem to have any connection with the true "oxytropic" irritability of the root tip.